The arid lands: their use and abuse

THEMES IN RESOURCE MANAGEMENT
Edited by Professor Bruce Mitchell, University of Waterloo

R. L. HEATHCOTE

The arid lands:
their use and abuse

Longman
London and New York

To Caroline

Longman Group Limited
Longman House, Burnt Mill, Harlow
Essex CM20 2JE, England
Associated companies throughout the world

*Published in the United States of America
by Longman Inc., New York*

© Longman Group Limited 1983

First published 1983

British Library Cataloguing in Publication Data
Heathcote, R. L.
  The arid lands. – (Themes in resource management)
  1. Desert resources development
  I. Title   II. Series
  333.73   QN541.5.D4
  ISBN 0-582-30048-7

Library of Congress Cataloging in Publication Data
Heathcote, R. L.
  The arid lands.
  (Themes in resource management)
  Bibliography: p.
  Includes index.
  1. Arid regions. I. Title. II. Series.
GB611.H35, 1983   333.73   82-17935
ISBN 0-582-30048-7

Set in 10 pt Linotron 202 Times

Printed in Hong Kong by
Astros Printing Ltd

Contents

Contents

Contents

List of figures

List of plates

List of tables

Acknowledgements

Upon taking up an appointment in 1966 at Flinders University – a new university in the "driest State in the driest continent", it seemed appropriate to offer a course on the problems of resource management in the world's arid lands. This book owes much to the ideas which have been presented in that course over the following years, ideas shared with and often stimulated by the students themselves. Additional research for the book was undertaken with the help of funds from Flinders University and David Moss and Kathy Ahwan provided faithful research assistance. Mrs. Jean Lange typed the manuscript with cheerful speed and accuracy, while Andrew Little, Jens Smith and David Johnson drew the maps and diagrams with professional skill. My wife provided the index and many hours of constructive criticism and proof reading, while the Series Editor provided informed critical advice. To all of these people I give my sincere thanks.

Finally, I wish to extend my thanks to the various institutions, publishers and individuals who gave permission for the reproduction of various items in this book: with the exception of Plate 17.1 all photographs were taken by the author. The errors which remain are my responsibility.

R. L. Heathcote
Adelaide, February 1983

We are grateful to the following for permission to reproduce copyright material:

American Association for the Advancement of Science, Westview Press & the author, H. L. Shantz, for extracts, our Table 2.3 (adapted) & our fig 6.1 (adapted) from fig 1 p 7 (Shantz 1956); American Geographical Society for our fig 15.6 (adapted) from figs 1 & 4 (Melamid 1973); American Scientist for our figs 4.3a (modified) from fig 6 (Judson 1968) & 14.2 (modified) from fig 10 (Skinner 1976); Angus & Robertson for our fig 4.3b (modified) from fig 4.2 (Marshall 1973); Arab Communicators for our fig 14.1 (Belgrave 1975); The Architectural Press Ltd for our Table 5.3 (extract) (Adams & Willens 1978); Association of American Geographers for an extract & our Table 6.3 (Germsmehl 1976); Australian Academy of Technological Sciences for our Tables 5.7 & 5.8 (Modified) from Tables 1, 2 & 3 (Evans 1979); Australian Conservation Foundation for our Table 8.3; Cambridge University Press for an extract & our Table 4.5 (Goodall & Perry 1979); Clark University for our Table 11.1 (modified) from Table 1 (Crossley 1976) & our fig 17.1 (modified) from fig 1 (Hamming 1958); C.S.I.R.O. for our fig 10.2 (Griffin & Lendon 1979); The Ecologist for our fig 18.3 from fig 1 (Baker 1976b); Econ Verlagsgruppe for our fig 15.1 a & b (Schneider 1963); Elsevier Scientific Publishing Co for an extract & our Table 4.3, p 39 (Dregne 1976); W. H. Freeman & Co for our figs 7.1 (adapted) from *Scientific American* Feb 1970 (Love 1971) & 9.1c (adapted) from *Scientific American* Sept 1960 (Deevey 1971); the author, Harold F. Heady for our fig 7.2 (Heady 1975); The John Hopkins University Press for our Table 13.7 (extract) from Table 8 *Resources for the Future* (Rutton 1965); The Institute of British Geographers for our fig 5.2 from fig 2 (Amiran 1978); International Centre for Arid Land Studies for our Table 6.6 (modified) from *Arid Land Plant Resources* (Weihe et al 1979); International Committee for Irrigation & Drainage, Pakistan, for extracts & our Table 13.2 (extract) (Gulhati 1955); Dr. W. Junk Publishers for our fig 10.1 (modified) from fig 1 *Ecological Studies in Southern Africa*, Monographiae Bioligicae XIV (Tobias 1964); the author, R. B. Lewcock, Middle East Centre, Univ. Cambridge for our fig 15.1c (Kirkman 1976); McGraw-Hill Book Co for our fig 3.1 from pp 392, 394–5 (Lobeck 1939) & our Tables 7.4 & 8.1 (Watt 1973); Organisation for African Unity for an extract & our fig 7.3 (Kokot 1955); Pergamon Press Ltd for our Table 18.3 (extracts) (U.N. Secretariat 1977); Pergamon Press Inc for our fig 15.2c & d (modified) from figs 4.13 & 4. 14 (Golanyi 1979): Rand McNally & Co for our figs 3.1 (extract) & 9.4 from *Goodes World Atlas* © Rand McNally & Co; The Royal Geographical Society & the author, Dr. Peter Beaumont for our fig 5.1 (Beaumont 1974); The Royal Society & the author, R. F. Hadley for our Table 11.4 (extract) from Tables 2 & 3 (Hadley 1977); the author, Prof. B. S. Saini for our figs 15.2a & b from figs 2 & 10 (Saini 1962); Society for Social Responsibility in Science, Canberra for our Table 8.2b (extract) (Diesendorf 1979); Sydney University Press for our fig 11.1 (extract) from fig 12.3 *Australia: a Geography* (Heathcote 1977); United Nations University, Tokyo for our figs 10.4e (modified) & 15.3 (modified) (Cordes & Scholz 1980) & our fig 12.2 (modified) (Heathcote 1980); University of Arizona Press for our Table 13.8a & b from *An Arizona Case Study* (Kelso et al 1973); University of Chicago Press & the author, Professor D. Johnson for our fig 10.4a–d (Johnson 1969).

Foreword

The 'Themes in Resource Management' Series has several objectives. One is to identify and to examine substantive and enduring resource management and development problems. A second objective is to assess responses to these management and development problems by researchers and policy-makers. A third objective is to explore the way in which resource analysis, management and development might be made more complementary to one another. *The Arid Lands: their use and abuse*, the third book in the series, addresses all of these objectives.

Regardless of the definition of 'arid lands', Heathcote demonstrates that they are vitally important on a global scale. While they contain only about 15 per cent of the world's population, the arid lands account for a third of the world's land area, provide a fifth of the world's food supplies, produce over one-half of the world's precious and semi-precious metals, and contain the bulk of the world's oil and natural gas reserves. These attributes make the arid lands significant at political, economic and military levels. Events and activities in the arid lands often have ramifications throughout the rest of the world.

Heathcote, who has a long-standing interest and record of research in the arid lands, has produced a book which is rich with concepts and information. By using examples from a variety of world regions, he is able to identify and evaluate a broad range of problems, research responses and management strategies. As a result, his book should be of interest not only to students and resource managers, but also to others who are looking for information or guidance about the opportunities and difficulties being encountered in the arid lands.

The book is divided into four sections. The first section introduces the arid lands, and addresses the issue of defining and delineating them. In this context, Heathcote makes a useful distinction between arid and drought-prone areas. In the second section, the resources (water, plants, animals, energy) of the arid lands are examined. This review of resources is explicitly set against the background of the nature of aridity and arid terrains. The third section addresses the evolution and impact of human activity in the arid lands. Here, attention ranges from an overview of general resource use strategies to examination of specific resource use activities such as nomadism, ranching, irrigation agriculture, mining, and urban design and development. This section highlights the necessity for resource managers to interrelate problems in urban and rural areas. It also emphasizes the necessity of recognizing local conditions and customs prior to prescribing solutions to problems. The fourth section focuses upon fundamental problems and questions of resource management in arid lands, including land tenure, alternative resource-management systems, and the role of perceptions in shaping approaches to resource management.

An array of examples is used. Comparisons are drawn among experiences in the arid lands of the United States, Australia and South Africa. Other experiences, especially those in the Middle East and the arid regions of Africa, are carefully described and assessed. The broad range of experiences effectively challenges the tendency towards developing uniform solutions. Heathcote convincingly argues that a blend of traditional and contemporary resource management strategies is most likely to be effective in the long run.

This book represents a major accomplishment by an individual who has been associated closely with analysis and development of resource-management strategies for the arid lands. It is to be hoped that this book will help to widen the range of choice for resource management for this important area of the world.

Bruce Mitchell
University of Waterloo
Waterloo, Ontario

May 1982

A note on the sources

In the text specific sources are indicated by the author and date of publication, with page numbers if necessary, and these sources are listed alphabetically in the References.

Any study of the global arid lands must draw heavily upon the UNESCO Arid Zone Research Series (Appendix 1). In addition, for English language materials valuable general bibliographies can be found in Paylore and McGinnies (1969) and Templer (1978), with regional examples being Paylore (1966) and Logan (1969). A comprehensive bibliography on desertification prepared for the 23rd International Geographical Congress in Moscow (Paylore 1976) is currently being extended. Earlier congresses such as the 21st in India 1968 and the 22nd at Montreal in 1972 contained special meetings and publications on the arid lands (see Das Gupta 1971 and Adams and Helleiner 1972).

Most geographical journals contain information on the arid lands at irregular intervals but the *Arid Lands Newsletter* (University of Arizona, Tucson) and the *Journal of Arid Environments* (Academic Press, London) provide regular and specific coverage.

General scientific works such as McGinnies, Goldman, and Paylore (1968) and Petrov (1976) provide good reviews of the physical resources, while research into human resource use is covered by a variety of sources. Until it became out of print Hills (1966) was the most useful general study, only partially replaced by Nir (1974) for the semi-arid lands. Several publications of the American Association for the Advancement of Science provide useful symposia (White 1956; Hodge 1963; Dregne 1970) while the University of Arizona has provided equally valuable collections of symposia papers (McGinnies and Goldman 1969; McGinnies, Goldman and Paylore 1971; and Amiran and Wilson 1973). A useful glossary of descriptive terms used in the arid lands is provided in Kharin and Petrov (1975).

PART I

The context of arid resource management

CHAPTER 1

Introduction

The fascination of the arid lands

From earliest records the global arid lands have had a multiple fascination for mankind. In that fascination may be found elements of four separate but mutually related themes which are still relevant to any discussion of future resource management in these arid lands. Those themes are:

1. A recognition of the arid lands as one of the 'wonders of the world'. With the polar regions and the tropical forests, the arid lands form one of the world's major landscape components – a global ecosystem of interlocking but distinctive life forms and environments (Pl. 1.1). As one of the world's wonders the arid lands have continued to exert a fascination for outsiders – whether scientists studying the unique life forms and environmental adaptations, armchair travellers with glossy coffee-table travelogues to scan, or tourists anxious to see for themselves.

2. The arid lands have not only been recognized to have a distinctive physical environment but also to contain some of the most important and impressive evidences of past human occupation of the earth. From the pyramids of Giza to the Great Wall of China and the Lines of Nasca, these artefacts of past societies pose questions as to the origins, flowering and demise of the civilizations which designed and built them (Pl. 1.2). In particular, since many of those ruins are now found in what appear to be desert environments, they raise questions about possible changes in those environments over time and whether such changes might have been from natural or man-induced causes.

3. The contrasts between the ruins of past civilizations and the structures of current arid land societies are paralleled by contrasts in the contemporary arid land use between the fruitful and the unfruitful lands (Pl. 1.3). This contrast, often razor-sharp on the ground, between the 'desert and the sown' again poses daunting questions about the nature of the resources and the productivity of the land. What combination of cir-

cumstances can produce such sharp divisions in nature? What element is so vital that its presence or absence can be reflected plainly in satellite imagery at hundreds of kilometres distance? The answers have not merely to do with water, but concern the whole question of the fertility of the land, the trends of that fertility through time and the relevance of those trends to human use of the arid lands.

4. Concern for the productivity of the arid lands leads into the fourth theme which is the recognition that the arid lands form a frontier of knowledge. Not only do they pose the questions noted above, but in the evidence of their past and present resource use, they provide useful experience of the problems associated with transfers of resource use systems from one ecosystem to another and the transfer of technologies from one resource use system to another. Since the nineteenth century the industrialization and socialization processes established earlier in Europe have been transported around the globe, and part of that movement has been into the global arid lands (Pl. 1.4). The results have been massive transformation of both the arid lands and their societies. That transformation has relevance for the whole history of human settlement in 'new' lands and for the future of resource development systems in both old and new lands.

Plate 1.1 Dripping Springs, Organ Pipe National Monument, Arizona, USA, photographed 1969.
Wide spacing of low bushes (creosote-bush) and alignment along run-off drainage lines with single tall saguaro cactus stems and isolated bush-like cholla cactus is evident in centre, with taller palo verde tree forms in foreground. Typical desert range and basin terrain.

3

Plate 1.2 The Registan, Samarkand, USSR, photographed 1976. The complex was built upon the site of an original caravanserai and market and comprises three structures. On the left is the Ulughbek madrassah, built as a university AD 1417–20; on the right the Shir-Dor madrassah, built to extend the university function 1619–36, and in the centre rear is the Tillya-Kari madrassah, built 1647–60 as a mosque. One of the major centres of culture and learning in central Asia in the fifteenth to seventeenth centuries and now a major tourist attraction, Samarkand celebrated its 2500th anniversary as a city in 1970.

In discussing the resource use of the global arid lands, therefore, these four themes will provide a framework. In Part I the basic character of the arid ecosystem will be established. This will be the context for human manipulation of those elements of that ecosystem which have, over time, been considered as resources. Part II will examine the historical sequences of human resource use in the arid lands, paying particular attention to the longevity of human use of the area and the fluctuating productivity of the settled areas. Finally, Part III will review the systems of resource use which have been applied to the arid lands and Part IV will discuss the problems which have arisen in the attempts to manage the resources of the arid lands.

The evolving knowledge of the arid lands

To put the contemporary international knowledge of the arid lands into context it is useful to summarize the evolution of Western knowledge of and attitude to the arid lands, since the recent international knowledge has been much influenced by Western thinking.

Plate 1.3 The Sossus Vlei, Namibia, photographed 1975. An oasis amid the high dunes of the Namib Desert formed around the claypan in which the Tsauchab River terminates. Currently a scientific reserve which typifies the sharply contrasting ecosystems found in the arid lands.

Pre-1800

Prior to the nineteenth century Western European knowledge of the arid lands was scanty and for Africa and southwest Asia had not been significantly updated since Roman times. The arid lands, whether of southwestern North America, Patagonia or Chile as known to the Spanish, or Africa, southwest and central Asia as more generally known, were seen as hostile environments harbouring heathen populations. From the arid lands of central Asia and North Africa had come serious threats to the survival of European culture, threats only removed in 1492 with the Spanish defeat of the Moors before Granada and in 1683 with the Austrian defeat of the besieging Turks before Vienna. The limited knowledge gained from the medieval crusades, the experiences of the Christian pilgrims to the Holy Land and the Renaissance interest in Arabic geographers such as Ibu Khaldun (1332–1406) provided scanty and often highly speculative knowledge.

1800 to the 1930s

From the late eighteenth century onwards the rapid expansion of European geographical knowledge included most of the global arid lands. A Danish expedition to Arabia in 1761–7 marked the first of a series of 'scientific' explorations by some of the most famous European explorers. The subse-

5

Plate 1.4 San Xavier Mission and suburban Tucson, Arizona, USA, photographed 1979.

  The Spanish mission of San Xavier del Bac in foreground, founded 1700 and still functioning, lies 24 km south of the Indian village, later Spanish presidio, of Tucson. Mexican then Anglo-American control followed and Tucson became the territorial capital of Arizona 1867–77. A recreational resort, centre for movie film production, air force base, and university support a sprawling modern city of over 300 000 people.

quent exploits of Samuel Baker (1821–93), Richard Burton (1821–90), Charles Doughty (1843–1926), Sven Hedin (1865–1952), Alexander von Humboldt (1769–1859), David Livingstone (1813–73), Baron von Richthofen (1833–1905) and many others made the headlines of the European newspapers. The explorers were particularly impressed by the indigenous nomadic life styles and the ruins of prior civilizations which littered the remotest deserts. The stimulus of such discoveries to the study of history, languages and archaeology, as well as to geography, was immense.

   From the mid-nineteenth century onwards settlers of direct and indirect European origin were beginning to extend European influence beyond that established earlier by the Spanish in the American arid lands, and to invade southern Africa, central Asia and Australia. For such settlers the new knowledge was directly pertinent and led to substantial transfers of people, plants, animals and ideas.

Since the 1930s

Desertification?
International concern for the arid lands markedly increased from the 1930s

onwards as the result of a combination of circumstances. The initial stimulus came from the evidence of environmental deterioration in the semi-arid Great Plains of the USA, where pressures of continuous cropping associated with drought, high winds and the rural depression of the early 1930s combined to create the massive soil erosion which the media rapidly identified as the 'Dust Bowl'.

Although soil conservationists see this Dust Bowl as the major impetus toward official concern for global soil erosion problems, in fact prior to and parallel with that event there had been evidence of significant environmental deterioration elsewhere (Lowdermilk 1953). In the same Great Plains during droughts in the 1880s–1890s (Brown 1948), in eastern Australia from droughts in 1895–1902 (Heathcote 1965), in South Africa from droughts in 1918 and the early 1920s (Kokot 1955), and on the southern edge of the Sahara (the Sahel Region) in the mid-1930s (Jacks and Whyte 1938) the data accumulated. This evidence of the deterioration of the productivity of the land, certainly in the short term and, it was suspected also in the long term, led to massive official disaster relief and rehabilitation measures.

Further drought-triggered disasters occurred in southeastern Australia in the 1940s, South Africa in the 1950s and the USSR in the 1950s – the latter being soil erosion from the newly ploughed lands of the Virgin Lands scheme. In the early 1970s further droughts, environmental degradation and human suffering in the Sahel brought massive international aid and stimulated a global review in 1977 of the so-called 'desertification' process (UN 1977a and 1977b; UN Secretariat 1977).

Post-war reconstruction

To this concern for the expansion of the global deserts had been added the experiences of warfare in the North African deserts in 1940–2 when scientific knowledge of the terrain and climatic stresses had been applied to the military training (e.g. the German Afrika Korps). Indeed, the practical experience from the mechanized desert warfare had relevance to post-war off-road vehicle technology. In addition, at least one scientist's interest in the arid lands was sparked off by his participation in the fighting, and R. A. Bagnold's monograph on dune sands (1941) was in part a result of his wartime experiences as a tank commander.

Plans for post-war reconstruction implied the use of new technology to increase the global food production for what was recognized to be a rapidly expanding population. In this increase of production the arid lands were to play their part. One school of thought held that they offered new areas to be brought into production for the first time with the new mechanized equipment; the other cited the evidence of soil losses, abandoned farmlands, and the ruins of past civilizations as an alarm signal to begin the process of *reclaiming* land from the advancing deserts. Whichever argument was used, there was a common aim to make the 'desert bloom as the rose'.

7

Introduction

The UNESCO Arid Zone Research Programme
The most important international scientific effort along these lines was the UNESCO Arid Zone Research Programme which was set up in 1951 and ran until 1971. The aims were the collection of data on the global arid lands; the standardization of data-collection units and areas to allow international comparisons to be facilitated; the study of different systems of land uses around the arid lands; and the planning of future resource use in the arid lands. Arising out of the work undertaken as part of this programme came a series of publications – the Arid Zone Research Series (Appendix 1) which laid the foundations of global scientific knowledge of the arid lands.

The evangelists
Paralleling this scientific research was a series of popular studies whose titles alone convey the evangelistic fervour with which they were written. The *Rape of the Earth* was the title used by Jacks and Whyte in 1938 to describe the impact of soil erosion around the world – not least on the edge of the deserts. They stressed the worldwide scope of the problem and the relatively small size of the remedial measures in progress. In 1949 Paul B. Sears' book *Deserts on the March*, described the 'shameful waste of our [USA] natural resources' from the war effort and the reduction of the marginal semi-arid agricultural lands to deserts. Its message according to the publisher's dust-jacket had 'a Biblical simplicity that fits well the greatness of its theme'. Two years later the science correspondent for the British newspaper the *News Chronicle*, Ritchie Calder, provided a survey of the work being undertaken by the *Men against the Desert* (1951). It reported on the 'battle' being waged against the advancing sands and in the words of the first Chairman of the Food and Agriculture Organization of the United Nations, Lord Boyd-Orr, it 'fires the imagination and provokes the conscience'.

In 1955 the Israeli author A. Reifenberg provided his account of *The Struggle between the Desert and the Sown*, where the efforts of the new nation of Israel to reclaim the deserts by a combination of applied scientific research, innovative technology, a nationally co-ordinated effort and sheer hard work, were set out as the culmination of a long history of desert advances and retreats from earliest records onwards. In 1966 a retired British forester (R. St Barbe Baker) with experience in Kenya and Nigeria and founder of the 'Men of the Trees' Society, provided his scenario for the *Sahara Conquest*. Not surprisingly, the desert was to be reclaimed by massive tree-planting schemes to establish both wind-breaks and soil stability across the arid areas. The amelioration of conditions thus resulting would lead, he thought, to significant increases in potential food production to help feed the growing world population.

To this series of documentary accounts we should add Frank Herbert's science-fiction saga *Dune* (1965), which, dedicated to the dry land ecologists, explored the ultimate logic of a rainless planet and the trials and adaptations of its plant, animal, and human populations. More recently John Updike's

8

novel *The Coup* (1979) has portrayed the predicament of one of the newly independent arid nations of the African Sahel.

The new arid nations

For some countries, particularly those newly independent nations on the fringes of the Sahara, interest in the arid lands is vital. For several the whole of their area is classed as arid (Table 1.1) and their economic future is dependent upon the adequacy with which they adapt their development strategies to the stress of their arid environments. That such nations have a voice and a vote at the United Nations helps them to bring their problems

Table 1.1 The arid nations

Type		Number (%)	Percentage of nation arid or semi-arid (%)	Nations (* = independent post-1945)
Group	Description			
I	Core	11 (16.7)	100	Bahrain*, Djibouti*, Egypt, Kuwait, Mauritania*, Oman*, Qatar*, United Arab Emirates*, Saudi Arabia, Somalia*, South Yemen*
II	Predominantly arid	23 (34.9)	75–99	Afghanistan, Algeria*, Australia, Botswana*, Cape Verde*, Chad*, Iran, Iraq, Israel*, Jordan, Kenya*, Libya*, Mali*, Morocco*, Namibia*, Niger*, North Yemen*, Pakistan*, Senegal*, Sudan*, Syria, Tunisia*, Upper Volta*
III	Substantially arid	5 (7.6)	50–74	Argentina, Ethiopia, Mongolia, South Africa, Turkey
IV	Semi-arid	9 (13.6)	25–49	Angola*, Bolivia, Chile, China, India*, Mexico, Tanzania*, Togo*, USA
V	Peripherally arid	18 (27.2)	<25	Benin*, Brazil, Canada, Central African Republic*, Ecuador, Ghana*, Lebanon, Lesotho*, Madagascar, Mozambique*, Nigeria*, Paraguay, Peru, Sri Lanka*, USSR, Venezuela, Zambia*, Zimbabwe*

Note: Total = 66 nations. Group I to III form the 39 basically arid nations; Group IV are the 9 semi-arid nations, and Group V those 18 nations where aridity is significant only at a regional level within the nation.
Source: Paylore and Greenwell 1979: 17–18.

into the international arena, particularly when those problems pose a massive human tragedy as in the Sahel famine of 1968–74. There is little doubt that the 'visibility' of this particular famine, in an area where droughts and associated famines had been common previously, was enhanced because the nations could raise their independent voices in concert at the United Nations. The same famine in the same area in the 1930s would have been seen as an internal matter for the British, French and Italian colonial powers and would probably have not come to the attention of the international media let alone the international political community.

Indeed, the emergence since 1945 of newly independent arid land nations has brought new political significance to North Africa and southwest Asia (Table 1.1). With Israel fighting for its existence against the majority of Arab nations; with the Sahel nations suffering internal unrest and civil wars as a result of both super-power and regional political rivalries; and with the North African nations jealously watching each other's political alliances, there are fears that a future global conflict might begin here.

The energy crisis
To this political importance derived from their newly independent and often unstable status, and from their mendicant character as receivers of international aid, many of the arid land nations have added the power which has come from their control of the main petroleum oil supplies of the world. The so-called 'energy crisis' of 1973, created by the sudden increase of oil prices set by the producers (the Oil Producing and Exporting Countries – OPEC), brought a new political significance to the arid lands. For the first time those who were major oil producers were able to dictate the price of their product to the consumers, who were mainly the Western industrialized nations. Given the then increasing dominance of oil as the major fossil fuel, it appeared that the global political power base might be beginning to shift from the traditional location in the temperate humid lands to the subtropical arid lands. In fact this has not proved to be so, but there is no doubt of the continuing political importance of the arid nations of North Africa, Arabia, and southwest Asia.

While fossil fuels continue to dominate the energy systems of the industrialized world, the search for the alternatives has included solar power and the massive supplies available in the arid lands have come under consideration. Even the export of surplus solar energy, for example as hydrogen, has been considered.

Aridity in space?
Finally, interest in the global arid lands has come from an unexpected quarter – namely research into outer space. Evidence from the Viking series of space probes has shown that 'aridity' is a feature of the planet Mars, where surfaces seem to be dominated by a rubble of recognizable chemicals

10

and minerals (such as iron, silicon, sulphur, calcium, etc.) but no water, an atmosphere 95 per cent carbon dioxide, 3 per cent nitrogen, and only 0.3 per cent oxygen, and temperatures ranging diurnally from −30 °C to −85 °C (Carr 1980, Young 1976). In the universe, aridity appears to be the norm – Frank Herbert's *Dune* was remarkably perceptive.

CHAPTER 2

Defining the arid lands

The problem

Any examination of the literature on the world's arid lands demonstrates that there are many different definitions of the area referred to, and many different definitions of the aridity which is suggested to be the distinguishing characteristic. The differences stem partly from the fact that the criteria on which the definitions are based refer to the deficiency of a resource (basically moisture) in the face of demands for that resource. Because demands will vary, for example between those of an urban water engineer designing a city's permanent water supply and those of a nomadic pastoralist searching for temporary water for his livestock, the thresholds of significance for the aridity characteristics will vary and the definition itself will vary as a result. Further differences in definition may result from the interpretations of different scholars concerned for different significance thresholds for aridity. Finally, definitions may be complicated by semantic confusion over the evolution of the meaning of the term 'desert'.

There seem to be two separate types of definition of arid lands – the one a series of literary definitions based on the changing meanings of 'desert', the other a series of scientific attempts to establish a meaningful boundary to the arid zone. In discussing the use of the arid lands, the latter is more important than the former, but literary definitions are worth brief investigation as evidence both of differing appraisals of the arid lands through time and of the danger in relying upon descriptive terms divorced from their temporal context.

Deserts: the literary definitions

Although derived from the Latin root *desertus*, meaning solitary, the English word 'desert' has had a variety of meanings in the past and still can be interpreted in various ways at the present time. The meanings identified in

12

Table 2.1 show this variety over time.

Several characteristics have recurred over the period from the thirteenth century onwards. Concepts of *barren lands*, i.e. useless and uncultivated, run through the period up to the present time. Lack of *water* is mentioned at the outset, but not until the twentieth century were the specific dimensions

Table 2.1 Changing definitions of 'desert' (1225–1968).

1225	'a desolate, barren region, waterless and treeless, with but scanty growth of herbage'. J. A. H. Murray (ed.), *A New English Dictionary on Historical Principles*, Oxford, 1897.
1398	'Places of wodes and mountayns that be not sowen bea calleyd desertes' – any wild uninhabited region, including forest land. Murray, ibid.
1880	'(desertus – solitary), a term used to denote any portion of the earth's surface which, from its barrenness, as in the case of the arid plains of Northern Africa and Arabia, or from its rank exuberance, as in the case of the *Silvas* of South America, is unfitted to be the site of great commercial and industrial communities. Many names, each differing in meaning to some extent, are employed to designate the desert-plains of different countries. The Desert proper may be said to signify the vast sandy plains of Africa and Arabia; while the flats extending from the Black Sea on the north, and from Persia on the south, onward across Tibet and Tartary to the northeastern coasts of Asia, are called *Steppes*; those in the northern division of South America, *Silvas* or desert-forests; those in the other portions of South America, Llanos and Pampas; and those in North America, *Prairies* or *Savannahs*. All these, though widely differing in individual characteristics, have in common the important physical features of wide extent and uniform general level.' *Chamber's Encyclopaedia*.
1909	'A desert is a country with such an arid climate and such a scanty water supply that agriculture is impracticable and occupation is found possible only for a sparse population of pastoralists.' J. W. Gregory: *South Australia*, London.
1924	'A desert has become by definition not naked sand and rock but a place of small rainfall with a sparse and specialised plant and animal life.' I. Bowman: *Desert Trails of Atacama*, New York.
1968	1. (a) Archaic – 'a wild uninhabited and uncultivated tract; a desolate unoccupied plain or coast or pathless woodlands wilderness, waste.' (b) 'any of the formerly unsettled regions of the United States between the Mississippi River and the Rocky Mountains thought to be arid and uninhabitable.' 2. (a) 'a region in which the vegetation is so scanty as to be incapable of supporting any considerable population (as a region perpetually cold or covered with snow or ice or a region located in the interior of a continent by scanty rainfall especially of less than 10 inches annually.' (b) 'a more or less barren tract incapable of supporting any considerable population without an artificial water supply.' (c) 'an area of ocean believed to be devoid of marine life.' 3. 'a secluded place of worship for Huguenots, 1715–1802'. 4. 'a desolating or forbidding prospect (as from pathless emptiness, bleak unrelieved changelessness or monotony, futility of effort or destitution of mental or spiritual animation or stimulation)'. *Webster's Third International Dictionary*.

defined. Lack of *population* is also a continuing theme, some definitions suggesting the lands are uninhabited, others only sparsely inhabited, specifically by pastoralists. Reference to a distinctive *vegetation* is also frequent throughout the period, but the characteristics were variously given as both treeless *and* wooded, scanty, specialized, and xerophytic. The *landforms* could include both mountains and plains, but deserts also could be defined in terms of areas of *ocean* devoid of life and associated even with a Huguenot *religious retreat* of the eighteenth century!

Such variety illustrates both the virile nature of the language and its embellishment over time and, as we shall see later, may be used in the search for an understanding of how human appraisals of the arid lands have changed over time (Ch. 18). As a basis for a scientific study of the arid lands, however, the literary definitions of desert are inadequate.

Not surprisingly therefore, when the UNESCO Arid Zone Research Programme was launched in 1951, one of the first tasks was to provide a definition acceptable to governments and the global scientific fraternity. Several alternatives were available and these, together with that devised by Peveril Meigs as the basis for the UNESCO Programme, are discussed below.

The arid zone: the scientific definitions

A definition based on climatic data would not be the same as one based on soils, on vegetation, on animal distribution, or on land use.
(Shantz 1956: 3)

Definitions pre-1952

The scientific approaches to the definition of the arid lands have varied according to the aims of the enquiry. For Albrecht Penck's research into global landforms (1894), he needed to identify those areas of the world where evaporation exceeded precipitation and thus caused a major modification of the erosion processes. This boundary, or 'dry line' – comparable in some ways to the 'snowline' in mountainous areas – was never mapped because data on evaporation on a global scale were unavailable, but the concept was implicit in later climatic indices.

A different method was used by E. de Martonne and L. Aufrère in 1927 when they published a global map of the areas of interior basin drainage. The claim was made that because of inadequate precipitation large areas of the global land surface did not drain to the sea and the geomorphological characteristics of such areas were likely to be different from those areas which did drain surface waters to the oceans. They claimed that 'the deserts are "par excellence" the domain of interior-basin drainage, or endoreism' (Martonne and Aufrère 1927), and on the basis of their map suggested 31 per cent of the global land area was so affected.

For the soil scientists, a fundamental division in the global soils is between those whose input of precipitation is sufficient to flush out the soluble carbonates from the surface layers to depths usually beyond the reach of plant roots (the pedalfers) and those soils where precipitation is insufficient and where carbonates remain and often accumulate by evaporation of capillary-induced solutions at or close to the surface (the pedocals). Estimates of the relative area of these pedocals – as 'arid' soils – indicate approximately 43 per cent of the global land area (Shantz 1956: 6).

The climatologists faced by the problem of defining the arid lands, initially had to rely upon the work of the botanists since maps of world vegetation predated maps of global climates. Thus W. Köppen, whose global climatic classifications were gradually refined over the period 1900–31, had been trained as a botanist and used de Candolle's plant classification of 1874 as the basis for his climatic zones (1931), arguing that the natural vegetation reflected the model climate conditions. For de Candolle's xerophytes – those plants adjusted to dry conditions – Köppen devised his classification of the global dry climates where the temperature was always high enough for plant growth. On the basis of the presence or absence of a seasonal precipitation, he distinguished the deserts (his Bw climates) as 12 per cent and steppes or semi-deserts (his Bs climates) as 14.3 per cent of the land area, giving a total of 26.3 per cent of the land area with dry climates. However, the classification did not account for the variation in efficiency of precipitation between seasons. It remained for C. W. Thornthwaite (1948) to devise a series of indices to show the relationship between precipitation and evapotranspiration (the combined moisture loss from an area by evaporation and the transpiration of plants). His indices were based upon practical observations of these relationships for the summer rainfall for the semi-arid Great Plains of the USA, but were less successful when applied to areas without a summer rainfall. Nevertheless his methods did show the importance of attempting to calculate the effectiveness of precipitation for plant growth in arid areas and were to be adapted later by Meigs.

Meigs' arid homoclimates

Faced with the request from the United Nations for a map of the arid lands, Meigs drew upon and improved the previous studies. Introducing his definitions in 1953, he noted that the chosen criteria had to be ecologically significant, of world-wide application and suitable for mapping on a global scale. Not surprisingly, in view of the context of the United Nations' request and the concern for global food supplies, the classification was 'oriented toward agricultural potentialities' and rainfall and temperature were 'of outstanding importance' (Meigs 1953: 203–4).

Using Thornthwaite's method of calculating potential evapotranspiration, Meigs calculated a *moisture index* to represent the relationship between precipitation and evapotranspiration. Excluding the areas too cold for crop

15

growth (from Köppen's criteria) he then divided the moisture index according to the possibility of crop growth and added a further subdivision of the driest classification for those areas where no precipitation had been recorded for at least 12 consecutive months (Table 2.2). His calculations increased the

Table 2.2 Meigs' arid homoclimates

Moisture index[a]	Homoclimates[b]	Comment[c]	Annual average precipitation[d] (mm)
0 to −20	Sub-humid	Suitable for crops	>500
−20 to −40	Semi-arid	Suitable for some crops only. Includes natural grasslands.	200−500
−40 to −56	Arid	Not suitable for crops	25−200
−57 and below	Extreme arid	Not suitable for crops. At least 12 consecutive months with no precipitation and no seasonal precipitation	<25

[a] Relationship of precipitation to evapotranspiration.
[b] Climatic zones.
[c] Suitability of climatic zones for temperate grain crops without irrigation.
[d] Estimates.
Sources: [a–c] Meigs 1953; [d] Grove 1977.

Table 2.3 Estimates of global arid lands

Arid area	Botanic definition[a]			Climatic definitions (% world)				
	Vegetation type	Area million km²	% world	W. Köppen	C. W. Thornthwaite[b]	P. Meigs[b] 1952	1953	UN[c] 1977
Extreme arid	Desert	6.29	4.7	Bw−desert = 12.0	Arid = 15.33	4.0	4.3	5.8
Arid	Desert grass savannah	5.96	4.4			15.0	16.2	13.7
	Desert/grass shrub	27.45	20.4					
	Sub-total	33.41	24.8					
Semi-arid	Sclerophyll bushland	3.06	2.3	Bs−steppe = 14.3	Semi-arid = 15.24	14.6	15.8	13.3
	Thorn forest	0.88	0.6					
	Short grass	3.10	2.3					
	Sub-total	7.04	5.2					
Total		46.75	34.7	26.3	30.57	33.6	36.3	32.8

Sources: [a] Shantz 1956: 4−5; [b] Meigs 1952, 1953; [c] Rogers 1981.

global proportion of the arid lands over both Köppen's and Thornthwaite's figures (Table 2.3) and have remained a useful global estimate of the significance of aridity for the growth of temperate crops. Subsequently his threefold classification has been widely adopted and average precipitation figures suggested for the boundaries between the zones.

The arid lands defined

For the purposes of this study Meigs' definition of the global arid zone has been adopted (Fig. 2.1). In the context of a study of the use of the global arid land resources Shantz's argument for adopting the whole range of aridity as identified by Meigs has considerable logic:

> I would be inclined to include all the area from extremely arid to semi-arid on this belt [the arid lands]. In this whole range the only safe assumption is that every year may be extremely arid. The more humid years will take care of themselves, but the arid years set the pattern of use and must be anticipated and planned for if man is safely to utilize this area. Moreover, the methods of use best suited to arid lands are the safest means of utilizing the semi-arid lands.   (Shantz 1956: 3–4)

The case for adopting Meigs' definitions is further strengthened by the coincidence of the proportions of the globe defined as arid by the interior-basin drainage criteria and Shantz's estimates of vegetation cover (Table 2.3). The more recent United Nations map, *World Distribution of Arid Regions* (UN 1977c), has only minor differences with Meigs' original map (see Table 2.3) and the earlier map has been chosen to enable comparative data based upon it to be used in this book.

The pattern of arid lands shows the largest continuous area stretching from northern Africa through southwest Asia to central Asia, as virtually a barrier between northwestern Europe and southeastern Asia (Fig. 2.1). In fact Africa and Asia contain 37 per cent and 34 per cent respectively of the global arid lands. Outside this broad arid zone, Australia has 13 per cent of the global arid lands, North America 8 per cent, and South America 6 per cent, with a small area in Spain comprising the remainder.

Questions raised by the definitions

Comparison of the details of the various scientifically derived definitions of the arid lands raises some interesting questions relevant to past, and maybe even present, trends in resource use. Although the definitions based upon criteria of drainage, climate, and vegetation all generally indicate that approximately one-third of the world land area has arid characteristics, the 'arid' types of soils cover 43 per cent of the globe. Is this difference significant? If we assume, as do all soil scientists, that there is a general relationship between regional climates and regional soils then there would appear

17

Fig. 2.1   The arid lands and arid nations. *Source*: Meigs 1953

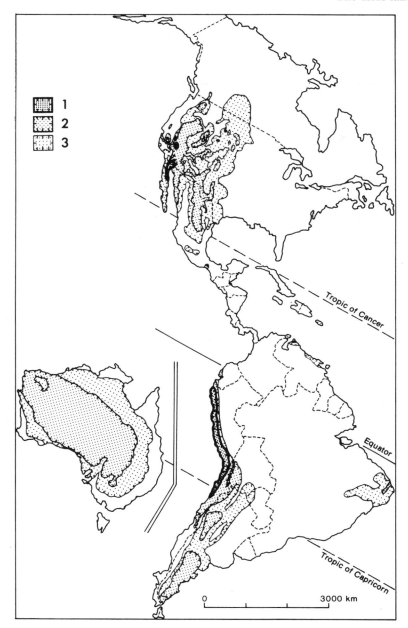

Key: 1 – Extreme arid areas; 2 – Arid areas; 3 – Semi-arid areas. National boundaries are indicated by dashed lines

19

to be areas of the world (some 7% or 10.2 million km²) which have soils which are more arid than would be expected from their climates. Is this true, and if so, does it reflect past misuse by man?

Another question is raised by detailed comparison of the climatic and botanical definitions. While there is agreement on the size of the extreme arid areas, there is considerable disagreement on the relative sizes of the arid and semi-arid areas. Again, if the constraints of regional climate have any relevance for patterns of regional vegetation, then the existence of large areas of the world which have an arid vegetation type but only a semi-arid climate constraint are difficult to explain. Table 2.3 in fact suggests that approximately 8.6 per cent of the world's vegetation by area (12.5 million km²) is in this way out of balance with its climate. In effect such areas appear to be poorer in quality than they ought to be, and again the question arises whether human activity may be responsible?

The arid nations

While climate knows no political boundaries, national climates vary considerably. Using Meigs' classification, it is possible to group the world's arid nations according to the proportion of their area which is arid or semi-arid (Table 1.1). For a 'core' group of 11 nations the whole of their territory is either arid or semi-arid, while another 23 'predominantly arid' nations have over three-quarters of their area affected by aridity, and 5 'substantially arid' nations have from a half to three-quarters of their area affected. For the remaining 27 nations, 9 fall into the 'semi-arid' group and 18 with less than a quarter of their area affected are only 'peripherally arid'. For some nations aridity is the dominant climatic characteristic, for others it is merely a regional condition.

The arid lands population

Amiran (in Hills 1966) estimated that the arid lands held not quite 13 per cent of the world's population (Table 2.4). By 1974 that share had only increased by 1 per cent, but the total (384 millions) had increased by almost two-thirds to 628 millions. Although the data may not be strictly comparable, my estimate for 1979 was a figure of 651 millions, some 15 per cent of the world's 4.3 billions.

Almost three-quarters of that population lived in the semi-arid zone with most of the remainder in the arid zone and few in the extremely arid zone. Over the period 1960–74 population in the arid and semi-arid zones increased by virtually two-thirds, whereas in the extremely arid zone the increase was by a fifth, but the relative proportions of population in each zone did not alter.

Table 2.4 Population in the arid lands

A. Arid lands population

Arid zone	Population				Increase 1960–74
	1960[a]		*1974*[b]		
	Million	(%)	Million	(%)	%
Extreme arid	5	1	6	1	+20
Arid	103	27	170	27	+65
Semi-arid	276	72	452	72	+64
Total	384	100	628	100	+63.5
% of world		12.8		14.0	+ 1.2

B. Arid nations population (1979)[c]

National groups		Population (*million*)	
Group	% arid	National	In arid lands
I Core	100	61.1	61.1
II Predominantly arid	75–99	305.2	221.0
III Substantially arid	50–74	128.2	47.0
Sub-total Groups I–III		494.5	329.1
IV Semi-arid	25–49	1 964.3	288.0
V Peripherally arid	<25	596.0	34.0
Sub-total Groups IV–V		2 560.3	322.0
Total arid nations		3 054.8	651.1
% of world		70.6%	15.0%

Notes and sources:
[a] Hills 1966: 219.
[b] UN Secretariat 1977: 268.
[c] Encyclopaedia Britannica, *Book of the Year 1981*, with arid population estimated from global population density maps in *Goode's World Atlas*, 15th edn, Rand McNally, Chicago, 1978.

The population in the arid lands then is a significant component of the global population, and while most live in the semi-arid fringes approximately 300 millions live in the arid heartlands – the arid and extremely arid zones.

PART II

The resources of the arid lands

CHAPTER 3

The nature of aridity

The climate is very intemperate, as in the middle of summer there are
terrible storms of thunder and lightning by which many people are
killed, and even there are great falls of snow and such tempests of cold
winds blow that sometimes people can hardly sit on horseback. In one
of these we had to throw ourselves down on the ground and could not
see through the prodigious dust. There are often showers of hail, and
sudden, intolerable heats followed by extreme cold.
(Father Carpini, Papal Legate to the Mongols, on the climate of the Gobi
Desert *c.* AD 1162. Quoted in Lamb 1963: 5)

Any attempt to understand the potentials for planning resource use in the
arid lands must first understand the nature of aridity itself. What are the
basic characteristics of the arid climates, what are the constraints they
impose upon life forms, how permanent are these characteristics and what
is their relevance to other aspects of the arid environment?

Aridity and drought

As Father Carpini noted in 1162 the climate of the arid lands can be very
variable. In discussing the nature of these climates it will be useful to distinguish between those factors which create a low absolute level of moisture
availability – in other words aridity, from those which create a temporary
and unexpected imbalance between the supply of, and demand for, moisture
at a specific time and place – namely drought. Aridity of itself does not imply
drought nor does drought only occur in arid climates. They are separate
although interrelated phenomena.

The causes of aridity

Aridity results from a combination of factors affecting the capacity of the meteorological conditions to supply moisture to an area. These factors include the basic physics of air movement, the global pattern of insolation and the geometry of land and sea relative to atmospheric movements.

The major subtropical arid lands lie in the subtropical high pressure zones where continually descending air is warmed and its capacity to absorb moisture thereby increased. These zones of continental tropical air masses are relatively stable both in their location and air mass characteristics. Their location is the result of the pattern of planetary atmospheric circulation, and the resultant associated hot arid climates appear to be permanent features of the contemporary global scene. Even the presence of large water bodies (such as the Red Sea and Persian Gulf) within the zone has no major regional impact on climatic conditions, the cooling effect of the water bodies being limited to within a few kilometres of the coast and the addition of moisture to the atmosphere being limited to within a few hundred metres of the high-water-mark (Pl. 3.1). Even permanently irrigated areas do not appear to modify the atmosphere in terms of temperature or moisture con-

Plate 3.1 Skeleton Coast, Namibia, photographed 1975. Within the narrow coastal corridor between the breakers of the southern Atlantic Ocean and the 40–50 m high crests of the mobile dunes of the Namib Desert runs the only railway and main road from Swakopmund (originally the only, and hazardous, port for German South West Africa) to Walvis Bay, originally a British whaling station and now the main port for Namibia. The fogs associated with the cooling of the hot air from the deserts by contact with cold air over the offshore Benguela Current are the only significant maritime influence on the local microclimate, and these together with the lack of safe harbours and treacherous sand bars offshore have led to frequent shipwrecks along this section of the Namibian coastline.

25

tent more than about 100 m downwind. In the face of such apparently stable air mass conditions proposals for weather modifications by flooding of arid basins by either sea-water or diverted river systems have no chance of significantly modifying the regional climate.

The basic physics of air movement also apply in 'rain-shadow' areas – the descending air on lee sides of high country, e.g. southern Patagonia, the Great Plains and parts of central Asia. Here, while the general atmospheric circulation might not be conducive to aridity, the local rain-shadow effects, if in operation for a significant portion of the year, create local arid areas. Again there is little that man can do to modify the atmospheric circulation patterns and such areas appear to be permanent arid lands.

In addition to this 'vertical geometry' of the land and its interaction with the atmospheric circulation, the pattern of land and sea also contributes to global aridity. The greater the distance from the ocean travelled by rain-bearing winds, the less the total moisture carried. The interior of the larger continents (Africa, Asia and Australia) therefore have less potential moisture available than the coasts.

Finally, the basic seasonal pattern of global climates fuelled by the solar oscillation between the tropics means that these areas receive the highest amounts of solar energy input. A map of monthly incoming radiation of more than 600 cal cm^{-2} day^{-1} shows the annual march of these areas through the subtropical arid lands north or south of the equator according to the waxing and waning of the summer conditions (Black 1956). Much of the massive energy which is represented by these figures is used in the evaporation of any moisture on land or in the atmosphere, but much remains to provide the high temperatures noted earlier.

The causes of drought

While the pattern of global aridity seems to be dictated by the basic global energy flux and the resultant patterns of atmospheric circulation, drought occurs as the result of specific shortfalls in moisture availability in the face of specific demands for moisture. Drought can therefore occur in any climate. While definitions of drought vary, a general definition is an unexpected shortage of available moisture sufficient to cause severe hardship to human resource use in the area so affected. An expected shortage, say from the effects of the seasonal 'dry' period, would not therefore be classified as a drought, but if the shortage occurred in the normally 'wet' season or the size of the shortage was significantly greater than normal in the 'dry' season (and had serious effects on the resource use) then a drought would be said to occur.

Given the spectrum of precipitation in the arid lands (from the extreme arid areas with none in any one year to the semi-arid areas with possibly a definite wet season) and the characteristic variability of precipitation over time and space, it is not surprising that droughts are recognized even in the arid areas. The occurrence of drought in many ways reflects the over-optimistic human appraisal of the moisture availability of an area as a component of its resource potential. Such appraisals have fluctuated over time with consequent significance for the success of resource use, as we shall see later.

The general characteristics of arid climates

Three components of the climates are examined here: (1) precipitation, (2) temperatures, (3) winds.

Precipitation

The range of absolute annual average precipitation in the arid lands, noted in Table 2.2 to be from *c*. 500 down to less than 25 mm, mask fluctuations which have vital significance for human occupation of the arid lands. At one end of the spectrum are the locations in the extreme arid areas like Arica (Chile) which over 17 years had an average precipitation of 0.5 mm and Cairo with an annual average of 28 mm which had precipitation in only 13 of the 30 years from 1890 to 1919 and then had 43 mm in one day in 1919 (Gautier 1970: 11). At the other end of the spectrum are the locations in the semi-arid areas where the average figures are larger and, while annual precipitation may fluctuate between less than half and over twice the average, there is always some precipitation in the year.

The fluctuation in precipitation in the arid lands means that the annual average figures are less useful here than perhaps anywhere else in the world. Because precipitation in the arid lands over a period of time, e.g. the meteorologically preferred 30-year sequence, does not usually show a normal (Gaussian) distribution, the mean and standard deviation are not particularly good measures of the spread of the data or of the probability of future precipitation occurrences (Lee 1979). A recent investigation of precipitation variability on a global scale showed that measured as average deviation in annual rainfall expressed as a percentage of the long-term average for 1931–60, the arid lands had the highest deviation – generally over 30 per cent (Morales 1977). Generally speaking, this deviation from the mean, or variability, increases as the absolute value of the mean decreases – the lower the rainfall the less chance of obtaining it in any one year. Given the trend, the predicament of places such as Arica and Cairo begins to make sense.

While fluctuations of precipitation well below the mean are a frequent and characteristic hazard for arid land use, fluctuations well above the mean are an almost equivalent hazard. For the subtropical arid areas in particular, that is, those areas on the edge of the major global monsoonal systems, an intrusion of these wind and air mass systems in the form of a tropical cyclone can bring intensive rainstorms with both positive and negative effects on the arid ecosystem.

On the positive side, the rainfall fills surface and subterranean water storages, breaks the drought, provides soil moisture for a vegetation response, and may even put out extensive bush fires. On the negative side, the massive inputs in a short time may cause major flooding, landslides and soil erosion, disruption of communications, buildings to collapse, and stimulate plague-proportion expansion of insect populations. Over 3 weeks in September–October 1969 heavy rains in Tunisia killed 542 people, left 100 000 homeless and caused damage to property of about $250 000. Cattle and sheep numbers were cut by 14 per cent and one watercourse was enlarged from 2.4 m wide to 24 m wide and 3 m deep. This event – defined as the 1 in 1000 year flood – brought the comment from the local peasants that for the first time in their memory 'water . . . behaved like a demon' (*Newsweek*, 22 December 1969). The occurrence of cyclone storms spreading into the Arabian and East African arid areas in 1948–9, into the Western Sahara 1968 and eastern Australia 1971–2, has been identified as the cause of locust outbreaks in the years immediately following (Rainey 1977).

The precipitation of the arid lands, while generally showing the lowest absolute amounts in the world:

> differs not only in its annual amount, but also in its spatial structure,
> spectrum of intensities, degree of uncertainty, etc.   (Sharon 1979)

While precipitation is usually absent for most of the time in the arid lands, when it does occur, whether from the intrusion of a tropical cyclone or local convectional heating, the rare downpour is likely to be heavy and localized. Even on a regional basis it is possible to talk of a seasonal rhythm of rainfall only on the semi-arid fringes and even then at the risk of considerable wishful thinking. Rainfall in the arid lands is infrequent and highly irregular, but when it does occur it is a very significant event.

Temperature

> The only common element is great range, both annual and daily. In the low latitudes, the difference between the coolest and the warmest months amounts to 15° to 30 °F (about 8° to 17 °C); in the middle latitudes, the yearly range usually amounts to 40° to 50 °F (22° to 28 °C).   (Cloudsley-Thompson 1969: 2)

Inadequate precipitation is matched by excessively high temperatures in the arid lands. The highest recorded temperatures at ground level have been recorded here (Azizia, Libya, 58 °C and Death Valley, USA, 56.7 °C) and summer months' mean maxima over 40 °C are common. Winter months' mean maxima may reach 30 °C in the Sahara, but the interior arid areas of

Asia maxima may be around 0 °C. In these latter areas the winter frosts are as strong a stress as the summer heat, but for most of the arid lands it is the excessive summer temperatures which form the main stress for all life forms.

The stress of high temperatures is modified by local ground surface conditions, the diurnal range and amount of water vapour in the atmosphere. Taking the latter first, the general lack of water vapour means that relative humidities (the difference between wet and dry bulb temperatures) are very low in the arid lands. For the central Sahara a range of from 4 per cent to 21 per cent was suggested by Gautier (1970). The lack of water vapour and associated cloud formation means both maximum heating from insolation during sunlight hours and maximum heat loss from radiation at night. Hence, the diurnal range of temperatures is high, often between 14 °C and 17 °C, but can be much higher:

> The oasis of In-Salah in the central Sahara has had a range of 100 °F (55.5 °C) within a 24-hour period: from 126 °F to 26 °F (52.2° to −3.3 °C) (Cloudsley-Thompson 1969: 2)

Such daily fluctuations place enormous physiological stresses upon all life forms.

The continuously high solar insolation in the low latitude arid lands (some 80% of the potential available) produces high energy fluxes. Energy budgets for arid areas are relatively few, but those which have been produced illus-

Table 3.1 Energy balance and resultant temperatures at Bukhara Oasis (USSR)

Criteria	Surface conditions	
	Bare ground	Cotton field
A. Energy balance (cal cm^{-2} h^{-1})		
Energy input	125	124
Energy absorbed	−79	−57
Energy radiated	46	67
Radiated energy used:		
– to heat air	−36	−3
– for evapotranspiration	0	−60
– to heat soil	−10	−4
	−46	−67
B. Resultant temperatures (°C)		
At 150 cm above ground	36	31
20 cm above ground	38	31
At ground surface	53	?
5 cm below surface	40	35

Note: Data for midday in August.   *Source*: Aizenshtat 1966

29

trate the range of conditions resulting from the high temperatures. Table 3.1 shows the protective effect of the foliage of the cotton plants upon soil and air temperatures near the ground, and illustrates the amount of energy absorbed in photosynthesis in plants when water is in virtually unlimited supply.

Wind

The low-latitude arid lands lie astride the subtropical high pressure systems which means that there are some areas of the hot, dry continental air masses which generate the seasonal trade winds towards the equator and the westerly wind systems towards the poles. At the heart of the high pressure system, in the case of the African, Arabian, and Australian arid lands however, the intensity and frequency of winds is not high by world standards. A map of world average wind speeds shows that the arid lands as a whole tend to average speeds of $c.$ km h^{-1} – 14 km h^{-1} and they do not as a result show up as major areas where wind power could be adapted as an energy source. On Eldridge's map, the arid lands generally appear to have annually less than 2250 kWh kW^{-1} output for wind machines rated at 40 km h^{-1} whereas coastal areas, especially at high latitudes, reach values up to 5000 kWh kW^{-1} (Eldridge 1976).

This overall view should not obscure the significance of both seasonal and intermittent winds in the arid lands. The winds function as moulders of the landscape, as carriers of soil particles and dust in sandstorms, as ventilators of buildings, as creators of psychologically stressful atmospheric conditions when their relative humidity is low and static electricity high, as energy source for remote water-pumps, electricity generators, food-grinders, and sailing vessels along the arid coasts. In all these cases the winds of the arid lands are a significant component of the climate as a resource and hazard and have been long recognized and identified as such by the local people.

Summary

The combination of high solar energy input, high temperatures and lack of moisture available for life support are the dominant characteristics of the arid lands climates which are relevant for human resource use. Dividing the arid lands on the basis of their seasonal temperature regimes, Meigs (1952) suggested four types: hot, mild, cool winter, and cold winter zones (Table 3.2). The largest proportion of the arid lands fell into the hot category; a quarter into the cold winter zone where winter temperatures would seriously restrict vegetation growth; and almost equivalent proportions into the mild and cool winter zones. Given that human ingenuity so far has found it easier to supply deficiencies of water than to supply deficiencies of seasonal temperatures adequate for plant growth, it is the hot and mild areas of the arid lands which might be seen to hold out the greatest hope of intensification of resource use.

Table 3.2 Climates of the arid lands

Type	Description		Percentage of arid lands
	Coldest months (°C)	Warmest months (°C)	
Hot	10–30	>30	43
Mild	10–20	10–30	18
Cool winter	0–10	10–30	15
Cold winter	<0	10–30	24

Source: Meigs 1952.

Micro-climates in the arid lands

Within the broadly defined characteristics of the arid lands climates, significant modification occurs as micro-climates. These are the climatic conditions
affecting small highly localized environments which might be as small as the
shadow of a pebble, or a rodent's burrow, or as large as a coastal strip only
a few metres wide but several kilometres long. Usually however, as one
researcher suggested, it is the 'climate near the ground' (Geiger 1965).

The significance of these micro-climates for human resource use is that
they usually offer less arid conditions for plant and animal life. The three
aspects of the micro-climate where modifications are most beneficial are in
relative humidities of the air, reduction of extreme air temperatures and
wind speeds.

Modification of relative humidities

The main modification is of air temperatures by nocturnal radiation,
shade effects or mixing with cooler air masses (as along coasts). The
results of the drop in air temperatures increase the relative humidities of the
air and the chances of moisture condensation either as fogs or dewfall.
Although research on such characteristics is scanty, enough is available to
suggest that such sources of condensation can provide significant moisture
supplies for limited areas and for limited periods of time. Thus the fogs along
the coast of the Namib Desert in South West Africa are thought to contribute about 150 mm of moisture per year to a narrow coastal strip. In southern
Israel dewfall is estimated to provide about 30 mm of moisture per year
spread over approximately 200 nights of the year. The absolute amounts of
moisture are not large, but they may represent a vital component to the total
water balance of plants in the area and may also sustain micro-organisms or
insects essential to the local ecosystem.

Reduction of temperature extremes

Any shade-giving object produces direct reduction of air temperatures in the arid lands because of the importance of the direct radiation component in the cloud-free atmosphere. Contrasts in air temperatures at midday between shaded and non-shaded locations can be of the order of at least 10 °C and obviously can mean the difference between life or death from heat stress in plants and animals. As we shall see in the discussion of plant and animal resources, the creation or use of existing micro-climates is a fundamental survival strategy for such life forms, both to reduce the high midday temperatures and to buffer against the low night temperatures.

Reduction of wind speeds

Shelter from wind movement reduces the amount of moisture loss by evaporation from plants and soil surfaces and also reduces the cooling effect of such evaporation. For the colder arid lands, man-made or natural wind-breaks hold snow-fall for longer periods than unprotected surfaces and this snow may be a useful source of water. Estimates of the area affected by protection from the wind suggest that a downwind area whose length is up to twenty-four times the height of the wind-break may have significantly reduced wind speeds. The optimal protection comes from barriers which are 40–50 per cent permeable (zero permeability causes damaging turbulence) and wind-protected wheat yields have been claimed to increase by 47 kg ha^{-1} on the semi-arid Great Plains. Where vegetation is used as the wind-break, however, its competition for soil moisture may reduce crop yields close to the barrier where soil moisture is the critical factor. For the resource planner, the problem is to weigh the costs of the wind-break (in terms of space used which is therefore no longer productive, its use of soil moisture, installation and maintenance) against the increased returns from the protected areas. Recent research has suggested that the economic benefits may not always be greater than those costs (Tinus 1976).

Summary: significance of micro-climates

The characteristics of the macro-climates of the arid lands are considerably modified by the factors which create micro-climates. The micro-climates appear to be particularly important as niches within the general arid lands whence life forms can retreat for survival or even live out their life cycles. Generally, the higher and more rugged the terrain, in terms of relative relief, the greater the role of micro-climates. Thus the mountains offer maximum modifications of the arid climates, not only by the reduction of air temperatures with height above sea level (about 1 °C per 102 m) but by shade effects and air drainage among the peaks and valleys. In his study of winter climates in the mountains of southern California, Logan noted differences in shade/sun air temperatures of up to 12 °F (6 °C) and rapid decrease of air temperatures after sunset (about 22 °F or 11 °C in one hour). He commented:

In summary, strong air drainage, direct insolational heating, lack of retention of heat by the air, and rapid nocturnal radiational cooling are the major elements that determine winter temperatures in this mid-latitude desert mountain and basin area. The end result is an extremely varied set of micro-climates, which in turn is strongly reflected in an extremely varied group of vegetation associations and animal habitats.

(Logan 1961a: 252)

In his use of plant and animal resources as well as his own habitats, man has made specific use of existing, and created his own, micro-climates in order to survive in the arid lands.

Changing arid climates

Through the coalescing research of geologists, geomorphologists, archeologists, historians and climatologists, the last two decades have provided increasing evidence of past changes in the patterns of global climates. Although of interest for its own sake, such evidence is of immediate relevance to resource use in the arid lands in two ways. First, any study of contemporary resource use in the arid lands needs to know if the lands have always been arid, or if not what were the past climatic conditions and can any long-term trends be distinguished? Second, any attempt to plan future resource use in the arid lands needs to know to what extent the current arid climates will be maintained in the future and the trend of any possible changes in those conditions. To answer the first question we need to ask the age of the present extent of arid climates and the significance for any present and future resource use of past non-arid conditions; to answer the second we need to ask how long the current arid climates can be expected to last relatively unchanged.

The age of the contemporary arid climates

The debate over the age of the contemporary global climatic patterns is complex, based upon evidence of varying quality and subject to various interpretations. That there have been significant changes in the past is not disputed. However, the nature, speed, extent, and timing of those changes *is* disputed. The best global review of the evidence is in Lamb (1977), but for the arid lands two studies by Grove (1977 and 1978) are particularly useful.

Reviewing the climatic changes since 20 000 years BP (before the present) Grove suggested that in terms of biological activity three periods could be distinguished. From at least 20 000 to approximately 13 000 BP was a period of low biological activity and reduced vegetation cover:

Ice desert occupied most of the surface within 40° of the Poles; loess

33

formations were accumulating in middle latitude, semi-arid regions of the present day, and although there were extensive lakes in southwest United States of America and other mid-latitude regions now semi-arid, they were the result of lower temperatures and evaporation rather than increased precipitation. . . . In general, the inter-tropical regions were drier than at present. . . . Active dunes occupied much of the Sudan zone of Africa . . . the semi-arid regions of north-west India . . . North-Central Australia . . . and savanna lands in the basin of the upper Orinoco and San Francisco rivers of South America. Arid conditions thus extended into regions now semi-arid.   (Grove 1977: 460–1)

From 13 000 to 7000 BP temperatures and precipitation increased gradually but irregularly, leading to a period of high biological activity and lusher vegetation cover. With temperatures comparable to the present but up to 150 per cent higher precipitation, extensive lakes formed in the present semi-arid areas as ancestors of the present Lake Chad on the edge of the Sahara, on the Okavango River in the Kalahari Desert, in an enlarged Caspian Sea, and Lake Eyre in Australia. The area of arid lands at this time was very limited and the semi-arid lands:

in Africa and possibly over much of the globe were limited to certain west coastal strips in low latitudes and to regions that are now arid.   (Grove 1977: 462)

From 7000 BP onwards a period of declining temperatures and somewhat lower precipitation brought conditions up to the present.

A short answer to our initial question as to the age of the present arid climate seems to be, on a global scale at least, that the present arid conditions are less than 7000 years old. What does this mean for human resource use in the arid lands? Three relevant areas might be identified.

The relevance of the age of the arid climates

First, it is obvious from geomorphological research that several landforms in the current arid lands reflect wetter climatic processes. This is true for relict glacial landforms – cwms and terminal moraines for example in the arid mountains of Morocco (Awad 1963) – which provide a spectacular relief in no way related to contemporary arid erosion processes. Of more immediate practical concern, however, is the realization that several of the soils in the arid lands owe their characteristics to previously different climatic regimes. Grove (1977: 464) comments:

In typical semi-arid regions, soils derived from dunes that were active in the last glacial period, though leached and even gullied in the succeeding period of high biological activity of Early Holocene times, are productive under cultivation . . . In the basins of Chad and on the plains east of the White Nile in Sudan . . . soils derived from riverine,

lagoonal and lacustrine clays [of the Early Holocene wetter period], are
of increasing agricultural importance . . . The clay soils of the Gezira
irrigation scheme in the Sudan are derived from sediments of the Blue
Nile laid down at this time of high discharge in the Early Holocene.
Both the dune and the lacustrine clay soils are more fertile than older
soils in the same general regions, which have been strongly leached
over a long period of time and commonly include lateritic ironstone.

In these cases the soils owe their agricultural potential to the past *non-arid*
climate.

Similar positive potentials are found in the areas of loess formed in the
earlier glacial period, which now comprise some of the best grain-producing
areas of the semi-arid lands (Fig. 3.1). However, in some cases the oscillations
of wetter and drier conditions led to the accumulation of salts or lime-rich
layers (caliche or calcrete) in the soil and these may prove barriers to moisture percolation and root growth (see also Ch. 4).

A second consideration is the suggestion from the geologists that the earlier more humid conditions led to the accumulation of large amounts of surface run-off in aquifers which are no longer receiving water inputs. In many
cases this water predates by many thousands of years the last major humid
climate in the present arid lands, but there is no doubt that the finite water
resource such aquifers offer is a direct result of previous non-arid climates.
As we shall see, this resource has been particularly significant in arid land
resource use over the last two decades.

Third, the relatively recent history (on the geological time-scale) of the
non-arid conditions means that for several parts of the present arid lands the
remote ancestors of the present inhabitants may have experienced substantial changes of climate which required significant changes of resource use for
them to survive. For eastern Africa the period of high biological activity was
associated with fishing peoples operating on the extensive lakes of the interior. The drying up of these lakes from 5000 to 4500 BP resulted in pastoralists abandoning traditional, but now desiccated, grazing areas. Grove
(1978: 295) suggests that this retreat:

> may have stimulated the settlement of the Nile valley for irrigation and
> accelerated the domestication in the Sahel zone [the southern edge of
> the Sahara] of several cereals that are now widely cultivated.

On the global scale such local adaptations must have been going on at varying rates of success throughout the last 7000 years, and the inhabitants of
the arid lands at AD 1500, before the European discovery of the 'new
worlds' of the Americas and later Australasia, must have represented the
survivors of this process of adaptation to environmental stress.

Not only the human inhabitants of the arid lands must have adapted themselves to the changing conditions. For the plants and animals the stresses
must have been considerable. The very evidence of the fluctuating nature

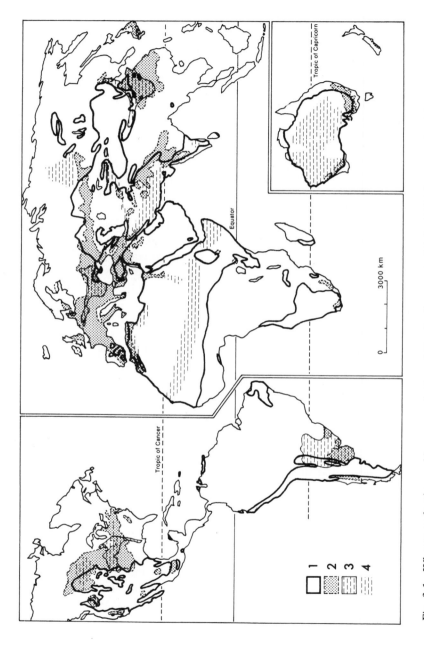

Fig. 3.1 Wheat production and loess areas in the arid lands. *Sources*: 1 – Meigs 1953; 2 – generalized from E. P. Espenshade (ed.) (1978) *Goode's World Atlas*, Rand McNally, Chicago, p. 30; 3 and 4 – Lobeck 1939. *Key*: 1 – Arid lands; 2 – Wheat production areas; 3 – Loess dominant; 4 – Loess present.

of those stresses over the last 20 000 years – a mere phase in the long history of the evolution of life on earth – might suggest that since the ecosystems of these areas have suffered in the not so remote past fluctuations of climate greater than those currently experienced, they must possess an inherent ability to cope with the current stresses. The oscillations of the past climates may have produced a resilience in the current ecosystems that would enable them to survive the stresses of current climatic fluctuations.

An unchanging arid climate?

Despite fears in 1972 of a major global climatic change, fears stimulated by the 1972 failure of grain crops around the world and the culmination of five years of drought in the Sahel, most climatologists now see the fluctuations which did occur as merely short-term aberrations in the long-term relatively stable pattern of climate. Grove (1977) suggested that in the last 200 years we had experienced on the global scale the range of climatic conditions which had been operating over the last 3000–4000 years. In other words, no major shift of conditions could be documented.

This finding should not blind us to the possibility of significant oscillations in regional climatic conditions over several years. In the semi-arid areas of eastern Australia, for example, Gentilli (1971) has shown that for the three 30-year periods from 1881 to 1970 the climate varied, being in 1881–1910 significantly wetter, in 1911–40 drier, and 1941–70 wetter again. This was associated with a displacement of the semi-arid/arid boundary by some 100 km. The period of the drier conditions 1911–40 was one of officially sponsored attempts to intensify land settlement in the semi-arid areas of New South Wales. Not surprisingly, the attempt to put more people on the land at a time when its productivity potential was in fact declining because of increasing aridity had limited success (Heathcote 1975).

Studies of trends in recent African climates have suggested that 'runs' of conditions significantly different from the current averages have been experienced within the period of scientific records. There have been sequences of up to 8 years in duration where conditions were much wetter than average, and for drier conditions sequences of up to 6 years' duration (Morales 1977). Thus, while massive changes in the global energy fluxes would appear to be necessary for catastrophic changes in climates to take place, there is evidence that seasonal oscillations with significant localized impacts on resource use *are* quite likely and can occur at any time.

Human modification of arid climates: actual and potential

While the bulk of this book discusses the methods by which man has adapted to the arid climates of the world, discussion of arid land resource use must also consider the possibilities for modification of arid climates. Such modification needs to be considered at two levels – first the macro-climate or regional level and second that of the micro-climate. While there is abundant evidence of human modification of the micro-climates in arid lands, evidence of an ability to modify macro-climates is less certain.

Modifications of the global arid climate

The role of human activity in modifying (accidentally or deliberately) the world climate has become a contentious issue among scientists in the last decade, partly as a result of the coalescence of evidence of climatic changes around the world from various disciplines, and partly because of the concern for future global food supplies at times of significant alterations in climatic patterns such as the droughts in the early 1970s.

Commenting in 1973 that the prior 2 years of harvest failures around the world had shown the vulnerability of the global food supply, the Director-General of the Food and Agriculture Organization (FAO), A. H. Beerma, suggested:

> despite all our technological progress . . . harvests are still far too often at the mercy of the weather. In this respect, at least, man has so far failed to master his natural environment.
>
> (Quoted in Newman and Pickett 1974: 877)

The implication was that only when man had mastered his environment would the harvests be assured. How far off is mastery of the global climate?

The short answer is, a long way off. The global energy fluxes are so great and so complex that human intervention to manipulate regional climates appears to be beyond the current or foreseeable technological capacities – even if such modification as implied by Beerma were considered vital to the future of man on earth.

Such a short answer, however, should not hide the fact that there is evidence of human impact on the global climates. The problem is that the precise results of this impact are not clear, and the impact is the accidental result of human activity which was not intended to modify the global climate. The most obvious example is the debate about the role of the increasing amounts of carbon dioxide (CO_2) being fed into the atmosphere as the by-products of motor vehicles and factory use of fossil fuels and the burning of vegetation as part of agricultural practices around the world. There appears to be no doubt that the CO_2 content of the atmosphere is increasing, vehicle emissions alone are increasing at 7 per cent per year (Watt 1973: 72), but the problem is that the scientists cannot agree on the effects this increase will have on the global climate. Most favour the so-called 'greenhouse' effect, where the CO_2 in the atmosphere will absorb solar long-wave radiation in the air and prevent short-wave reradiation from the earth being diffused out of the atmosphere. The result is a gradual heating of the air of the earth's surface and a warming of the globe (Kellogg 1978). This would increase the potential growing season in the mid-latitude grain areas of the world (including the fringe of the semi-arid lands), but although it might increase the precipitation the scientists are not sure where this would happen. If the analogy with previous more humid periods is any guide, the increase might be within the arid lands while the polar areas would be drier. If so:

> the milder climate at high latitudes and the likelihood of wetter conditions in parts of the sub-tropics suggest that it might be possible to grow more food for the growing world population, even without advances in agricultural technology.
> (Kellogg 1978: 78)

This would be the optimistic view both of the location and the effects of the climatic changes resulting from CO_2 increases alone.

A complication which causes the greatest debate among the scientists is the role of other additions to the atmosphere, particularly hydrocarbon aerosol propellants and dust, whether from wind-borne soil or volcanic materials. Davitaya's (1969) study of dust layers deposited on central Asian glaciers established a chronology of varying seasonal deposits from 1793 to 1962. The maximum accumulation occurred at the same time as the volcanic explosion of Krakatoa in 1883 and for several years thereafter the thickness of seasonal deposits was three times the average. Significantly, the seasonal thicknesses increased past the 1920s up to ten times the average, the result, according to Davitaya, of the combined effects of the spread of cultivation in the 1930s into the Russian steppes, the destructive effects of the Second World War, industrial pollution, and the massive plough-up of the steppes in the 1950s as part of the Soviet 'Virgin Lands' development scheme. Apart from the documentation of the impact of human activity on the dust content of the atmosphere, the relevance of the findings was that he estimated air temperatures during major dust storms – such as had deposited these layers – had been increased by up to 10 °C, with associated increased evaporation of any available moisture and desiccation of plant and animal life. His findings have been borne out by research into the increasing dust storms experienced over the Thar Desert, which has confirmed the increased temperatures and desiccating efforts of such phenomena (Bryson and Baerries 1967). In the arid southwest of the USA, however, a distinction has been made between the impacts of low levels and high levels of dust: the first increased air temperatures, but increasing dustiness apparently caused eventual reductions in temperatures. Further, the high-altitude dust particles were found to reduce the amount of solar radiation filtering to the earth's surface and so in fact would reduce lower-level air temperatures. The study's conclusion posed the as-yet-unanswered problem:

> We foresee the possibility of future volcanic activity as tending to reduce surface temperatures; while we view the increased activities of man in producing industrial and agricultural pollution and the enhanced activity of wind in raising dust as tending to warm the surface. Both

> factors must be considered in assessing the likely course of the earth's future climate.   (Brazel and Idso 1979: 437)

The accidental impacts of man on the global climate are difficult to establish, not least because they may be masked by other factors creating change.

The uncertainties as to human impact on the global climate have not prevented many elaborate designs being put forward for direct human manipulation of the climatic patterns. Most of these plans are scaled-up versions of the observed effects on the micro-climates of human activity.

Modification of local arid climates

Human activity in the arid lands can both increase or decrease the local aridity. Increases in the aridity will result from removal of protective vegetation, thus allowing increased heating of the ground surface. Increases in the albedo (reflectivity of the ground surface) by salting of the soil will lead to increased air temperatures from increases in reflected radiation. Decreases in aridity will result from introduction of water into the area by irrigation and cloud seeding. Reduction of wind speed (and hence drying effect) results as we have seen above from the introduction of windbreaks.

Given the demonstrated ability of such practices to change the local aridity, if only temporarily, the schemes to modify larger areas for longer periods have been basically larger versions on the same principles. As reviewed by Glantz (1977) those relevant to the arid lands included:

1. the creation of artificial lakes in the arid areas by diversion of surface flows from humid areas;
2. the cloud-seeding of the Inter-tropical Monsoon Zone to encourage a wider impact into the arid lands:
3. the creation of a sequence of tree shelter-belts across the Sahara; and
4. the laying of massive asphalt surfaces in the desert to increase surface temperatures and create thermal updraught to create clouds and rainstorms.

As yet, the costs and uncertain results of such schemes make them highly unlikely strategies for resource improvement (Katz and Glantz 1979).

As we shall see in the examination of human use of the arid lands, the local arid conditions have been significantly modified already by man. Perhaps a better strategy for arid resource use is to recognize the variability of the global climates, particularly in the arid lands, and organize resource use to compensate for the inevitable shortages of production in one location by transfers of the equally inevitable surpluses from other locations. Behind such a strategy is the belief, borne out by the history of the global climatic variations, that:

> Climate in a general way, obeys the law of conservation. That is,

whenever one large area is getting too little precipitation, another is getting too much.   (Newman and Pickett 1974: 880)

Such transfers are technically possible, but given the current state of global politics, highly unlikely.

CHAPTER 4

The arid terrains

Within the areas dominated by the arid climates lie a variety of terrains offering a variety of resources for human manipulation. It is worth noting the general characteristics of those terrains before examining their detailed patterns over the globe, the associated soils, and the processes which are continually modifying them and thus directly or indirectly modifying their resource potentials.

General terrain characteristics

The popular arid lands' image derived probably from film stereotypes is of endless empty shifting sands dotted with the occasional oasis of luxuriant plant and animal life. While large areas of the arid lands would conform to this description, there are larger areas with very different characteristics.

The traveller in the arid lands is conscious that the lack or sparse nature of vegetation enables him to see clearly the rock structures which are the bases for the landforms (Pl. 4.1). Indeed it was the ease with which these structures could be identified in the arid southwestern USA in the 1870s and 1880s which inspired geologists and geomorphologists such as C. Dutton and G. K. Gilbert to provide theories of the origin and processes affecting these landforms, and gave a boost to the scientific study of geology. The mile-deep exposures of the strata from the Palaeozoic period onwards in the walls of the Grand Canyon of the Colorado River for example were a Rosetta Stone for the unravelling of the geological history of not only the southwestern USA but for global geology also (Rabbitt 1979).

The traveller is also aware of the angularity of those landforms and their dominant colour range at the red, brown, yellow end of the colour spectrum. The angularity, resulting from the infrequent but intensive nature of the arid land erosion processes, combined with the vivid colours of the rocks and soils, resulting from the unleached iron oxides and salts still present at the

Plate 4.1 Vegetation patterns in the Namib Desert, photographed 1975.
Vertically dipping strata of variable hardness have weathered unevenly,
thus providing uneven potentials for soil formation and the establishment
of a 'banded' vegetation. Scene is about 43 km east of Walvis Bay on the
road to Windhoek, Namibia.

surface, produce landscapes of striking beauty – as the expansion of the
tourist industry in the arid lands has shown.

For the precise nature of the resources offered by the arid terrains, however, we need to classify the variety in some meaningful way.

The patterns of terrains

On a global scale the arid lands include a range of geological structures, from
rugged 'Alpine' mountain systems formed since the Juraissic period (130
million years BP) to broad tablelands and plateaux, remnants of Gondwana
Shields of over 1500 million BP, and lowlands of relatively recent sedimentary
rocks often less than 1–2 million years old. The details of the distribution
of these areas show considerable variation (Fig. 4.1 and Table 4.1).

For the arid lands as a whole, the largest area is underlain by sedimentary
structures which contain both live and 'fossil' aquifers and fossil fuel deposits
(mainly of petroleum and natural gas but also coal). In terms of their potential for underground water and fossil fuels as energy sources, almost 40 per
cent of the arid lands are well endowed. The next largest unit, the Gondwana Shield areas (over a third of the total area) is much less endowed with
underground water potential or fossil fuels but may contain areas of precious

43

Fig. 4.1 Arid land terrains. *Source*: Murphy 1968

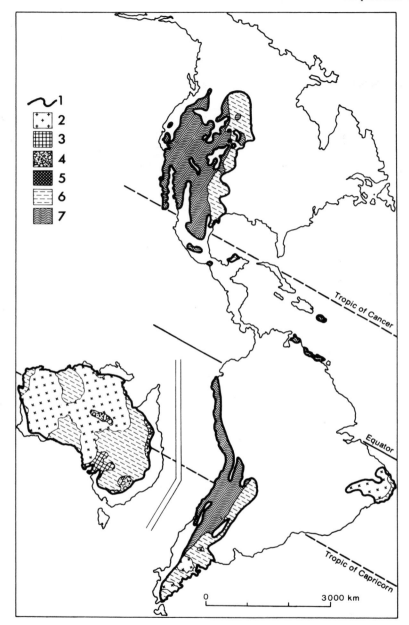

Key: 1 – Edge of arid lands; 2 – Gondwana Shields; 3 – Rifted Shield areas; 4 – Caledonian and Hercynian Remnants; 5 – Isolated volcanic areas; 6 – Sedimentary areas (outside Shields) 7 – Alpine systems

45

Table 4.1 Terrains of the global arid lands

Continent	Landform structures (% arid area)[a]					
	Alpine	Hercynian remnants	Gondwana Shields	Rifted Shield	Sedimentary areas	Isolated volcanic areas
Australia	0	2.1	61.7	3.1	33.1	0
Africa–North	0	0	56.3	7.4	35.4	0.9
–South	0	0	91.4	0	8.6	0
America–North	71.4	0	0	0	28.6	0
–South	52.7	0	4.0	0	43.3	0
Asia	30.5	7.2	4.7	4.7	52.9	0
Total arid lands	18.9	2.4	35.0	4.5	38.9	0.3

[a] Percentage of continental arid area covered by the landforms. The structures are: *Alpine* – mountain chains formed since Jurassic period; *Hercynian* – remains of mountain chains in Paleozoic/Mesozoic eras, relatively unmodified since; *Gondwana* – remnants of Gondwanaland, stable massive Pre-Cambrian rocks; *Rifted Shield* – block faulted areas (horsts and grabens); *Sedimentary* – post Cretaceous era deposits; *Isolated volcanic* – active or extinct volcanoes outside above areas.
Sources: Calculated from Murphy 1968.

and semi-precious metals. The mountainous Alpine and Rifted Shield areas offer the maximum potential for modification of the general arid climates and sites for possible storage for surface water in their rugged relief (Pl. 4.2). Only a small total area is covered by Hercynian structures with their potential for coal deposits, and the isolated volcanic structures are insignificant on a global scale but locally provide striking landforms.

Between the continents the characteristics of the arid lands vary further. The arid areas in North and South America are dominated by the Alpine mountainous areas with the remaining portion of sedimentary rocks – the Great Plains of the north and the Patagonian plains in the south. The Asian arid lands also have almost a third of their area in mountainous country, with just over half in the sedimentary plains of the central Asian steppes and the low tablelands of eastern Arabia. By contrast, the African and Australian arid lands have no major mountainous areas except for the small rift-affected areas of East Africa and South Australia. They are dominated by the ancient Gondwana Shield surfaces (from over 50% in Australia to over 90% in southern Africa), with sedimentary areas making up the bulk of the remainder. On a continental scale, therefore, the terrain-derived resources vary significantly around the world.

More specific evidence of the variation in the resource base afforded by the contrasts in terrain types can be found in the military classification of 'desert surface conditions' evolved for the southwestern USA, but applied

Plate 4.2 Arches National Monument, Utah, USA, photographed 1960. From the arid basin where the eroded natural arches in the massive Entrada sandstone have been protected as a scientific reserve the visitor can see the snow-capped peaks of the La Sal Mountains in the background. Apart from the aesthetic attractions of the reserve and the view, the snow-melt provides the waters for exotic rivers which are tapped for irrigation in the lower arid valleys. The area was created a National Park in 1971.

also to other arid lands (Table 4.2). Here ten different landforms have been distinguished in terms of their relative relief, slopes, and surface texture, and their relative areas in five arid areas calculated.

The results of this closer analysis of the terrains shows the importance of the mountain areas in all the 'deserts'. Dunes cover approximately a quarter of the area, but are up to one-third in Australia and virtually absent from the southwestern USA, where alluvial fans and bajadas form almost one-third of the surfaces. The lack of dunes and dominance of the fans reflects the importance of the mountainous terrain of the southern Rocky Mountains and the range and basic topography of the Great Basin, which themselves reflect the younger geological age of the landforms compared with those of the shield-dominated area of Africa and Australia (Table 4.1). The combined categories of desert flats and bedrock fields cover usually about a fifth of the arid land surfaces, but in Australia the proportion rises to almost one-third.

The relevance of these more detailed classifications lies in the information they provide on potential accessibility, presence or absence of soils and surface or underground water. Thus, if we combine the desert mountains with the 'badlands' and dune areas for the two African and one Australian

Table 4.2 Landforms of the arid lands

Landform type[a]	Arid areas (areas as % total)[b]				
	Southwest USA	Saharan Desert	Libyan Desert	Arabian Desert	Australia
A. *Playa*: 'nearly flat, sun-baked expanse of clay and salt', run-on area; periodic water cover; good access when dry; zero vegetation Alt. names: salina, claypan, takyr	1.1	1	1	1	1 (playa)
B. *Desert flat*: relatively flat, borders playa; relief 0.3–1.6 m, may include dunes up to 5 m high; slope 1 in 352; watercourses cross; vegetation sparse; access good	20.5	10	18	16	18 (stony desert)
C. *Bedrock fields* (i) *Pediment*: 'slightly inclined rock surfaces thinly veneered with fluvial gravels'; slope 2°–7°; surface coarse sand to boulders; vegetation on watercourses; access varies: good to poor (ii) *Desert dome*: 'convex surfaces with uniform and smooth slopes'; 4–10 km diameter; 180–700 m height; slopes 1°–4°; most vegetation types here. (iii) *Hamada*: 'bare rock of low relief' vegetation zero or sparse; access poor to fair	0.7	10	6	1	14 (shield desert)
D. *Regions bordering through-flowing rivers = canyonlands*: area eroded by tributaries to main stream; terraced surfaces; pebbles to alluvial	1.2	1	3	1	?

Table 4.2 (contd.)

Landform type[a]	Arid areas (area as % total)[b]				
	Southwest USA	Saharan Desert	Libyan Desert	Arabian Desert	Australia
surfaces; includes some badlands; maximum vegetation here; access poor to good					
E. *Alluvial fans and bajadas*: deltas individual or coalesced (bajada); relative relief to 1 m, if dissected, to 18 m; constant slopes; detritus grades from gravel and boulders at apex to sand and silt at foot of slope; mud flows occur; vegetation sparse; access good generally	31.4	1	1	4	13 (clay plains and floodplains)
F. *Dunes*: hill of windblown material (clay or sand); asymmetrical section with steep slip-off slopes to 32°; windward slopes 15°–19°; sand movement bulk within 2 m of surface, 90% within 0.3 m of surface. Several types reflecting in their geometry wind directions, barchan (crescent shaped), transverse and linear (parallel but separate), lunette (Australian, dune on lee side of playa, of gypsum from dried lake bed)	0.6	28	22	26	38 (sand desert)
G. *Dry Washes*: dry watercourse; U-shaped section; slope 2°–3°; sand, gravel, boulder bed;	3.6	1	1	1	?

Table 4.2 (contd.)

Landform type[a]	Arid areas (area as % total)[b]				
	Southwest USA	Saharan Desert	Libyan Desert	Arabian Desert	Australia
vegetation good; access for four-wheel drive					
H. *Badlands*: rough dissected soft sedimentary rocks; relative relief to 30 m; clay, silt surfaces; vegetation sparse; access poor	2.6	2	8	1	?
I. *Volcanic cones and fields*: recent volcanic surfaces; loose boulders, lava; slopes to 30°; vegetation zero; access poor	0.2	3	1	2	?
J. *Desert mountains*: bare rock masses; granitic (rounded); metamorphic (angular); sedimentary (canyons, amphitheatres)	38.1	43	39	47	16 (mountains)
	100	100	100	100	100

[a] From US Army Quartermaster Research and Development Center Techn. Rep. EP-53, '*A Study of Desert Surface Conditions*', 1957.

[b] Australian figures from Mabbutt 1976, others from source above. Mabbutt's classification is not strictly compatible and his classification descriptions are indicated in brackets.

'deserts', from 61 per cent to 73 per cent of the arid areas there pose considerable problems of access. Indeed, there is evidence which suggests that exploitation of the remoter arid lands has had to await the successful development of off-road vehicles, short-take-off aircraft and helicopters to cope with this accessibility problem. For areas with useful soil potentials we would look to the alluvial fans and bajadas, the dry washes and possibly the desert flats. Here also would be the best chance of underground water supplies.

Arid land soils

While the art of growing plants in chemical solutions without soil (hydroponics) is well developed, the costs involved prevent all but specialist uses for a basic food supply, and the people of the arid lands must rely upon and manipulate the existing soils as the medium in which the plants are to be grown. Yet knowledge of soils in the arid lands is much less than knowledge of soils in the humid areas of the world where the practical problems of agriculture have been of concern since at least Roman times. In Europe the so-called Agricultural Revolution of the eighteenth century encouraged enquiry into soil characteristics as a variable affecting plant growth and the discovery of the role of chemicals by Justus Liebig in the 1840s and the foundation of the Rothamsted Experimental Station in 1872 initiated longitudinal research into the role of chemical fertilizers and plant chemistry.

Research

> Detailed soil maps are not available for most of the arid regions. . . .
> In general, the more arid the climate, the less is the accuracy of the
> soils information. (Dregne 1976: v)

The expansion of European settlers into the semi-arid steppes of central Asia provided practical evidence of the role of climatic differences in soil characteristics. It is not surprising therefore, that two of the world's pioneer scientists were Russians – V. V. Dokuchaev (1846–1903) who in 1870 recognized that the soil profile was the key to a useful system of soil classification, and K. D. Glinka (1867–1927) whose classification on this principle in 1908 suggested that climate was the dominant factor in soil formation. His concepts of mature and immature soils were based upon the extent to which the soil-forming materials of rock and plants had produced a profile which reflected the influence of climatic factors – mature soils reflected the dominance of those factors, immature soils with incomplete profiles reflected the dominance of other factors such as waterlogging and the bedrock. For Glinka the arid lands had immature soils because the profiles were incomplete and often dominated by the underlying bedrock.

Paralleling Glinka's Russian work, E. W. Hilgard (1833–1916) began research at a California experimental station in 1875 on the semi-arid and arid soils as potential areas for agriculture. His publication, *Soils, their Function, Properties, Composition and Relations to Climate and Plant Growth in the Humid and Arid Regions* (1906) stressed the role of climate, particularly in its influence upon the depth at which materials leached out of the surface layers of the soil were deposited lower in the soil profile. In the humid areas he suggested this could be at depths well beyond the reach of plant roots, whereas in the arid areas the shallow penetration of the reduced amounts of precipitation meant that this zone of accumulation could be close to or even at the soil surface. Ideas based upon humid land experience had stressed that only the heavy clay soils were fertile, but Hilgard (noting that most arid land soils were light textured and the areas of heavier soil probably

51

reflected different past climatic influences) suggested that these light arid soils still contained many of the soluble chemicals useful to plants (lime and magnesium for example) which had been leached out of the upper levels of the humid soils. In contrast to soil scientists with experience only in the humid areas, he forecast a good agricultural future for the semi-arid soils of the steppes and natural grasslands which fringed, and in many areas included, substantial areas of arid land.

Coming as it did towards the end of the nineteenth century, the work of the Russians and Hilgard provided scientific support for a reassessment of the agricultural potentials of the arid areas, at a time when European settlers were beginning to push their agricultural systems out of the humid areas of Europe, Russia, the Americas, southern Africa, and Australia into the adjacent but relatively unknown semi-arid grass- and shrublands. Such research encouraged both the irrigators and the dry-farmers to press on down the rainfall gradients.

By the 1930s in both the USA and Australia the consequence of blind transfers of the humid land technologies to these semi-arid areas had become evident in the dust storms which blew topsoil from the eroding farmlands east to the national capitals. As a result, soil research was given a new stimulus and direction to prevent the loss of what was beginning to be seen as a national heritage. Despite the strenuous and often successful local work of the official soil conservationists (Bennett 1947), the *global* understanding of arid land soils was not markedly improved. The renewed concern for soil erosion in the 1970s as part of the desertification process highlighted the paucity of prior knowledge and showed how inadequately existing knowledge was applied. In fact, not until Dregne's book *Soils of the Arid Regions* appeared in 1976 was a comprehensive global view of the detailed characteristics of arid soils available.

Characteristics

Based upon the work of the American and Russian soil scientists the place of the arid soils in the global soil sequence can be illustrated by Fig. 4.2. Decreasing precipitation inputs and increasing evaporation from the soil surface areas means that the soluble salts remain close to the surface, while in areas of minimal precipitation they may be at the surface or only covered by wind-blown materials. Thus the arid soils generally are pedocals as opposed to the leached pedalfers. The lack of available moisture not only slows down the chemical processes but also the breakdown of plant materials into humus, so that the depth and quantity of humus in the arid soils is low and declines rapidly down the rainfall gradient.

Dregne's survey (1976) applied contemporary US soil survey research to the global arid soils and provided both classification and assessment of areas of the five major soil orders for the arid lands (Fig. 4.2 and Table 4.3). The two soil orders specifically arid in nature are the *Aridisols* with their high

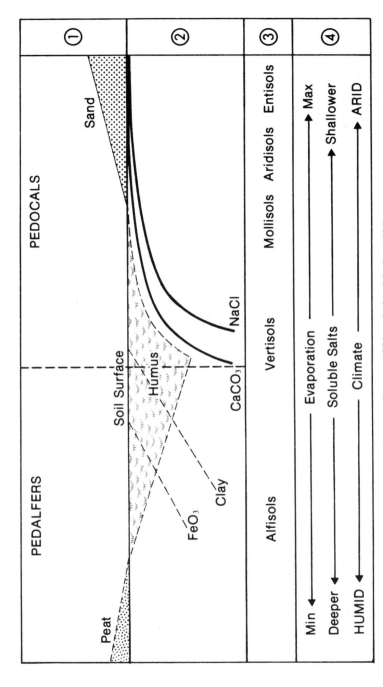

Fig. 4.2 Arid soils in the global context. *Sources:* Dregne 1976 and after Marbut 1935
The basic components of soil characteristics indicated on a spectrum from arid to humid climatic conditions
Key: 1 – Surface components; 2 – Subsurface components with soluble mineral accumulation layers indicated; 3
– Approximate location of new soil types (7th Approximation) on the spectrum; 4 Characteristic trends of eva-
poration and soluble salts as related to climate.

53

Table 4.3 Arid soils

Soil Order[a]	World arid lands			Continents (% of arid lands)				
	Area (million km²)	% of arid lands	% world land area	Africa	Asia[b]	Australia	North America	South America
1. Alfisols	3.1	6.6	2.1	11.8	—	7.0	3.8	13.4
2. Aridisols	16.6	35.9	11.3	27.7	41.7	44.2	44.8	27.9
3. Entisols	19.2	41.5	13.1	58.4	33.7	36.6	8.0	41.4
4. Mollisols	5.5	11.9	3.7	0.7	19.8	—	—	17.3
5. Vertisols	1.9	4.1	1.3	1.4	5.4	12.2	2.3	—
Totals	46.2	100%	31.5%	100%	100%	100%	100%	100%
Percentage of continent				59.2	33.0	82.1	18.0	16.2

Note: Soil descriptions are based upon the US Soil Conservation Service's 7th Approximation.
[a] Soil orders: Alfisols – 'ochric epipedon, an argillic horizon, moderate to high base saturation' plus moisture for 3 months' growing season. Aridisols – 'mineral soils of the arid regions . . . low organic matter content', salic/duripan horizon within 1 m of surface. Entisols – 'mineral soils' of no pedogenic horizons including alluvials and sands. Mollisols – dark surface, rich in calcium and magnesium, semi-arid and sub-humid grassland soils. Vertisols – 'deep and cracking clays of warm regions'.
[b] Excluding European Russia.
Source: Dregne 1976.

mineral/low organic matter content and hard-pan zone of mineral accumulations close to the surface, and *Entisols* – particularly the sandy types of undifferentiated soils. Between them these two orders cover some 77 per cent of the arid lands, ranging from 53 per cent of the arid lands of North America to 86 per cent of African arid lands.

Within the arid lands, however, Table 4.3 shows the presence of soils not specifically arid in character. Even *Alfisols*, the leached humid soils, are found in the arid lands of Africa and South America in particular, while Australia has 12 per cent of its arid lands with the deep cracking *gilgai* clays of the *Vertisols* whose origin is certainly non-arid. On the borders of the arid lands the semi-arid and nutrient-rich *Mollisols* form almost one-fifth of the arid lands in both South America and Asia.

One characteristic of the arid soils which is becoming more significant each year is the proportion of saline soils. Dregne's figures tend to hide this specific soil type, but the Russian researcher Kovda has provided some estimates of saline soils by nations (Table 4.4). While Kovda's estimates included non-arid areas of the nations, there can be little doubt that the core and predominantly arid nations contain almost two-thirds of the measured saline soil areas of the globe. Such are major problem areas for resource management.

Agricultural potentials

Life is in sands.
(A central Asian proverb quoted in Goodall and Perry 1979: 445)

As suggested in p. 52 arid soils do have significant agricultural potential, either because of the presence of relict soil types from prior non-arid periods or because of the presence of non-leached minerals within the potential root zone. Further, some terrains are likely to offer better soils than others. Using Table 4.2 we might forecast that useful soils will be found on the playas, desert flats, alluvial fans, dry washes and, possibly, the badlands. Depending upon the location, this would represent usually an area of from 14 per cent to 32 per cent of the arid lands with the USA having the largest area – 59 per cent. While not directly comparable with Table 4.3, such figures must be borne in mind when any attempt to assess the agricultural potentials of the arid lands is made.

The potential of the soils to capture, retain, and release moisture and nutrients to plants varies enormously, and perhaps surprisingly the central Asian proverb appears to be generally valid (Pl. 4.3). In the Negev (Table 4.5) and in South Australia (Table 4.6), except for the concentration of moisture in the watercourses, it is the sandy areas which offer the optimal regimes. A recent review suggested:

Sandy soils have the most favourable water regime [for plants]. They are wetted more deeply, have the largest store of available water, do

55

Table 4.4 Global salt-affected soils

Arid nations[a]	Area of Salt-affected Soils[b] (million km²)
Group I	0.2
Group II	4.4
Group III	1.0
Sub-total Group I–III	5.7
(% world total)	(62.8%)
Group IV	0.9
Group V	2.2
Total arid nations	8.7
(% world total)	(96.6)
World total	9.0
(% world arid lands)[c]	(19.5%)

Notes and sources:
[a] See Table 1.1.
[b] From Kovda noted in White 1978 (Table 6).
[c] Using Dregne's 1976 estimate of world arid land soils of 46.2 million km².

Plate 4.3 Sand forms and vegetation of the Kara Kum Desert, USSR, photographed 1976.

Complex mobile sand dunes are partially stabilized by small trees and shrubs of *Haloxylon* species, whose roots (as illustrated by the one held up for display) spread laterally distances several times greater than the height of the plant itself. The adjacent Repetek Sand Desert Reserve was created in 1928 to allow scientific study of the arid ecosystem and research results have been applied in the post-1950s reclamation of some of the Hungry Steppe and Kopet Dag areas to east and west respectively.

not lose moisture by surface and ground run-off, and have low evaporation. Thus, 80 per cent of the stored moisture is used in the transpiration of plants. The most unfavourable water regime is that of clay and loess plains . . . even the soils of rocky slopes have more favourable conditions. The natural stony mulch which apparently

Table 4.5 Soil:water regime in the Negev (Israel)

Habitat	Maximum water storage available in root zone		'Average' season water balance		
	Available moisture (mm)	Total moisture (mm)	Direct evaporation (mm)	Plant Use (transpiration)	
				Amount (mm)	% of total
Loessial wadi	300	300–800	50	250–500	83–63
Rocky slope	40	100	30	50	50
Loessial plain	40	100	45	35	35
Sand	50–60	120	30	90	75

Source: Goodall and Perry 1979: 443.

Table 4.6 Soil:water relationships in semi-arid South Australia

Water in soil[a]	Soil types					
	Coarse sand	Fine sand	Sandy loam	Loam	Clay loam	Clay
Field capacity (F)[b]	65	130	170	300	360	500
Wilting point (W)[c]	25	50	60	110	140	250
Available water (F–W)	40	80	110	190	220	250
(as % of F)	(61)	(61)	(64)	(63)	(61)	(50)
Precipitation needed for vegetation response[d]	25	50	60	110	140	250

[a] Points (× 0.25 mm) per 0.31 m of soil.
[b] Field capacity is that amount of moisture which a soil will retain before it begins to shed the excess. For loam soils this level is reached at about 19% by volume of water in the soil.
[c] Wilting point is that point in the reduction of moisture in the soil where the remaining moisture becomes unavailable for plant use. For loam soils this is about 10% by volume of water in soil.
[d] Assuming dry soil (zero water content), amounts in points (× 0.25 mm)

Source: Schulz 1967.

retards run-off, prevents the formation of a continuous crust, and slows the rate of evaporation. . . . On the other hand, the soils of clay and loess plains are covered by nearly solid crusts, poorly absorbing and rapidly evaporating moisture. Consequently, these soils lose much moisture unproductively. (Goodall and Perry 1979: 443–4)

In addition, the sandy soils respond most rapidly to the smaller precipitation inputs – the same rainfall might, for example, provide a pastoralist with a vegetation response for his livestock in sand-dunes but not be sufficient for any response on the clay plains.

The changing terrains

The resources offered by the arid terrains are constantly changing. Processes of weathering, erosion and deposition are constantly modifying the details of the patterns of landforms and associated soils.

On the global scale the rate of such changes seems to be at a maximum in the semi-arid areas, where the precipitation is still fairly high, but vegetation cover provides inadequate protection for the ground against erosive processes (Fig. 4.3A). Judson (1968) estimated (from the exposure of the roots of bristlecone pines some 4000 years old) that semi-arid areas with slopes of 5° lost up to 2 cm of surface material per 1000 years, and slopes of 30° lost up to 10 cm of material per 1000 years. He estimated that this 'natural' erosion peaked in areas of 300 mm average rainfall – the semi-arid country, and this is confirmed by Marshall (1973) for Australian conditions (Fig. 4.3B). Where mankind had disturbed the soil by cultivation, however, the rates of erosion were up to 100 times greater. Where excavations for buildings were made the rates could be up to 1000 times greater.

The natural processes of landform modification in the arid lands also reflect most of the processes in the humid lands, but the relative importance of individual processes varies. The geographer W. M. Davis in 1905 recognized that in the arid areas his earlier (1899) global theory of the Cycle of Erosion would have to be modified. The theory suggested that water erosion reduced youthful newly uplifted dissected terrain to old peneplain surfaces close to the base level of erosion, i.e. sea level. In the arid lands there might be no drainage to the sea (as de Martonne and Aufrère demonstrated in 1927) and the streams generally were shorter than the slopes. Therefore, there could be no universal base level and the lack of surface drainage systems together with the obvious role of the wind in transporting loose surface materials had to be recognized (Davis 1905).

The role of wind in eroding soft rocks, transporting sand and dust, and shaping dunes is obvious enough, and since 90 per cent of the materials are carried within 0.5 m of the surface this can pose problems for maintenance of buildings (especially paintwork), fence or power posts, road clearance, and silting of rail tracks and canals.

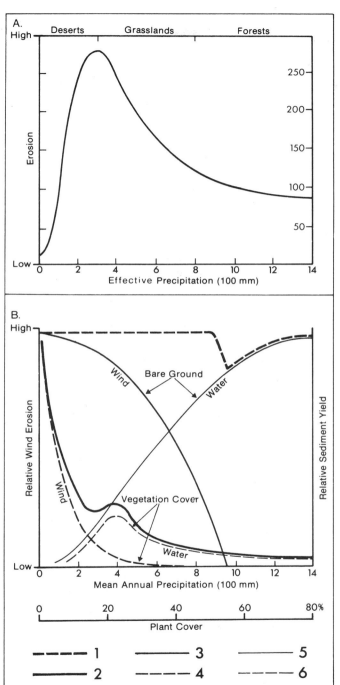

Key
(A) Average annual sediment yield (erosion) in relation to effective precipitation
Erosion – average annual sediment yield in tons per km^2 *Effective precipitation* – amount necessary to produce a given amount of run-off

(B) Water and wind erosion in relation to bare ground, vegetation cover and mean annual precipitation.
1 – Combined level of wind and water erosion from bare ground
2 – Combined level of wind and water erosion from vegetated surfaces with increasing cover
3 – Wind erosion from bare ground
4 – Wind erosion from vegetated surface
5 – Water erosion from bare ground
6 – Water erosion from vegetated surface

Fig. 4.3 Erosion in arid lands. *Sources*: (A) Judson 1968; (B) Marshall 1973.

Except for the dune areas the air traveller over the arid lands is constantly impressed by the evidence of surface-water drainage channels. Bearing in mind the supposed lack of precipitation the continued evidence of such watercourses is surprising. In part it reflects the high run-off from the extensive bare rock, sparsely vegetated or crusted surfaces in the arid lands, plus the torrential nature of some of the infrequent rainstorms. Water is a rare commodity in the arid lands, but when it does occur its capacity to modify the landscape can be enormous, as we have seen in Chapter 3. The impact of such transformations can be devastating. T. E. Lawrence described one such impact on the village of Wasta in the Wadi Safra in Arabia:

> A generation ago Wasta was populous (they said of a thousand houses); but one day there rolled a huge wall of water down Wadi Safra, the embankments of many palm-gardens were breached, and the palm-trees swept away. Some of the islands on which the houses had stood for centuries were submerged, and the mud houses melted back again into mud, killing or drowning the unfortunate slaves within. The men could have been replaced, and the trees, had the soil remained; but the gardens had been built up of earth carefully won from the normal freshets by years of labour, and this wave of water – eight feet deep, running in a race for three days – reduced the plots in its track to their primordial banks of stones.   (Lawrence 1963: 92)

Changes in the arid terrain could be sudden and devastating. To consider the terrains and associated soils as fixed resources in the arid lands is to deny both the sequence of natural erosion and deposition processes and the modification of those processes by the hand of man.

CHAPTER 5

Water resources

With 97 per cent of the world's water in the sea and 2 per cent in deep freeze, the world evidently is a fine place for whales and penguins, but it has its shortcomings for man.   (Nace 1969: 40)

The global distribution of water poses considerable problems for mankind in general. Those problems both of quantity and quality of water resources are exascerbated in the arid lands. The distribution of global water shows that even if the arid lands had a share of the land-based water equal to their share of the land area (excluding the ice-caps and glaciers), this proportion would only be 0.18–0.19 per cent of the total (Table 5.1). Due to the relative

Table 5.1 Water in the world

Types of water	% of total world water
Atmospheric vapour (water equivalent)	0.0001
World oceans	97.6
Water in land-areas:	
river channel storage	0.0001
freshwater lakes	0.0094
saline lakes; inland seas	0.0076
soil moisture; vadose moisture	0.0108
ground-water	0.5060
ice-caps and glaciers	1.9250
Total	100.0%
Water available to arid lands as percentage of that in land area (excluding ice-caps and glaciers):	
(i) if arid lands = 33.6% of world[a]	0.1794
(ii) if arid lands = 36.1% of world[b]	0.1927

[a] Meigs 1952.   [b] Meigs 1953.
Source: Chorley *et al.* 1969: 32

lack of freshwater lakes, the relative sparsity of soil moisture, and the relatively small (although still significant) local flows of the few major rivers in the arid lands, the proportion of global water is probably closer to 0.17 per cent. To assess the significance of this small but vital share to the arid lands we need to look at the location of those supplies in the arid lands, their quality, and the problems associated with their use.

Water in the arid lands

Surface waters in the arid lands can be divided into permanent and temporary supplies. These are relative rather than absolute terms since 'permanence' needs to be judged in terms of demand, which usually is increasing. The permanent supplies include the flowing freshwater rivers such as the Nile, Indus, and Yellow River in China, which have played a major role in global history as the sites for early civilization based upon irrigation. Yet the total average discharge of the major arid land rivers is only 6.6 per cent of that of the Amazon (Table 5.2).

Also included in the permanent supplies are the major inland seas and saline lakes, the bulk of which lie within the arid lands. Thus, 76 per cent of these waters are in the Caspian Sea and the bulk of the remainder lies within the Asian arid lands.

The temporary or ephemeral supplies include precipitation, whether as snow, rain, dew, or fog drip, which have been discussed previously in Chapter 3. The concentration of these inputs into the surface watercourses or as ground-water recharge provide important temporary resources, the manipulation of which is one of the skills of the arid lands populations.

Ground-water resources in the arid lands can be divided into archaic or living waters, depending upon whether the aquifer is being actively

Table 5.2 Major arid land rivers

River	Average discharge at mouth ($m^3 sec^{-1}$)
Indus	5 547
Nile	2 830
Missouri	1 953
Yellow (China)	1 500
Colorado	156[a]
Total	11 985
Amazon (for comparison as world's largest river)	181 120

[a] By 1979 irrigation usage had reduced the flow at the mouth to virtually zero.
Source: Chorley *et al*. 1969: 54

recharged. Innovations in drilling and pumping technology have made vast previously unknown or unobtainable ground-water supplies into resources over the last two decades, mainly as the by-product of the increased tempo of oil-search drilling. Indeed, it was the use of diamond-tipped rotary drills developed for oil-search purposes, which led to the discovery of the Great Artesian Basin in Australia in 1879, the most significant find of ground-water in the global arid lands prior to the Saharan discoveries of the 1960s. In the Great Artesian Basin the flows are from a living aquifer which is being recharged by rainfalls in the humid coastal ranges of Queensland; in the case of the largest Saharan discoveries, they are believed to be archaic waters approximately 25 000 years old, which are not being recharged. In both cases, however, use of the water is a 'mining' operation since the rate of use exceeds the replacement even in the Great Artesian Basin, where flows from more than 20 000 bores have fallen from the 1920s figure of about 3.2 billion litres per day to some 909 million litres per day currently. By AD 2010, the safe yield is expected to have fallen to 500 million litres per day.

The role of these underground supplies has proved to be vital to arid lands development, particularly from the latter half of the nineteenth century onwards. The role of such supplies, however, depended upon the quality of the resource.

Water quality

Water in the arid lands varies in quality from the purest drinking water to the salt-saturated brine of the Dead Sea. While the average drinking water in the UK and USA contains about 570 parts per million (ppm) of solids, and the World Health Organization sets an upper limit of 500 mg litre^{-1}, livestock in arid lands have been known to tolerate up to 3000 ppm and humans temporarily from 2500 to 4000 ppm. Variation in quality of supplies can be seasonal or due to unusually severe stress in the supplies. A drought and increased salinity of the Australian River Murray, which forms the main summer supplementary supply to the South Australian capital, Adelaide, led to the city's water reaching 900 ppm during the summer of 1967–8. The city is not in the arid zone, but its water supply passed through the zone and was affected by the water stresses there.

Tolerance of salts in water varies among different plants, but for most forms of life the presence of sodium chloride becomes critical at 0.5 per cent solution. By comparison ocean water has salt at 3.4 per cent and the Dead Sea at 27.5 per cent solutions. The range of salt tolerances of most domestic plants is known, but Boyko (1966) has shown that these tolerances can be extended by use of sandy soils and careful flushing of surplus salts (Table 5.3).

The salts which pose such problems for living organisms offer raw materials to industry. The salts of sodium, magnesium, and potassium, present

Table 5.3 Cropping potentials in saline soils

Salts in soil (NaCl ppm)	Ecological association	Potential crops
4000–8000	Suaeda fruticosa and Salicorwa fruticosa	Cabbage, beet, sweet clover. Limit for cotton, barley, oats and wheat
3000–5000	Suaeda fruticosa	Cotton, beet, artichoke, pomegranate, oats, lucerne
2000–3000	Suaeda fruticosa	Tomatoes, melons, marrows, olives
1000–2000	Atriplex halimus	Limit for potato, carrot, onions, pimento, peppers

Source: Adams and Willens 1978: 57 (after Chapman).

in the brackish waters of the arid lands, have been the basis for industries to produce those minerals. The Dead Sea waters were first used industrially to produce potash in 1930, and despite interruptions from warfare, solar evaporation from the salt pans is still producing potash, bromide, and magnesium chlorides, some for agricultural fertilizers in both Israel and Jordan. From the Dead Sea come life-giving nutrients for the soils of the arid lands.

Problems of water use

Ensuring the proper water quality is but one of the problems for water use in the arid lands. Four types of problems can be identified: first, access to supplies (whatever their quality); second, coping with poor-quality supplies; third, competing uses for the water; fourth, the potential for improving both quantity and quality by purifying processes.

Accessibility

The problem is that the water is not where it is wanted. The solutions, given that a supply to meet the demand is available somewhere, are technical ones relating to water transfer. Long-distance aqueducts date back at least to Roman times, but the increasing global trend towards urbanization has meant that cities in or fringing the arid lands have had to transfer drinking water long distances. In 1913 Los Angeles obtained the snowmelt run-off from the Sierra Nevada with a 373 km aqueduct and siphons. In the 1930s this was supplemented by a 426 km aqueduct from the Colorado River (Chorley *et al.*, 1969). In 1903 one of the then longest pipelines in the world was completed from reservoirs in the 'Mediterranean' climate of the Darling Ranges in Western Australia to bring drinking water some 562 km to the gold-mining town of Kalgoorlie in the arid interior.

These transfers were either by concrete-lined channels or enclosed pipe-lines so that the loss by evaporation along the route was relatively small. Transfers along unlined channels can be disastrously inefficient. Between 1925 and 1974 (when they were replaced by pipelines) a series of open unlined channels in northwestern Victoria (Australia) carried water pumped up from the River Murray to wheat–sheep farms for livestock and domestic water supplies. Investigations in the late 1960s discovered that less than 10 per cent of the water pumped out of the river actually reached the farms, the rest being either lost by evaporation or through seepage from the chan-nels, many of which crossed linear sand-dunes which traversed the farming country (Heathcote 1980).

Underground horizontal transfers were the traditional method for bajada or piedmont locations where an inclined shaft tapped aquifers by gravity for users downslope. Such tunnels – *qanats* or *foggaras* – were, and locally still are, vital to irrigation schemes in the mountainous arid lands of southwest Asia. However, current tunnelling is limited to cross-watershed transfers such as the Snowy Mountain Project in southeast Australia, where eastward-flowing rivers are diverted to supplement and maintain the irrigation capacity of the westward-flowing River Murray, and the Big Thompson Project of the western USA where Colorado River water is transferred eastward through the Rocky Mountains to the irrigation areas of the western Great Plains.

Vertical transfers by wells depended for their success upon the energy available for the vertical lift of the water. Before the use of fossil fuels and mechanical engines the maximum possible depth of wells seems to have been the 88.5 m deep Joseph's Well in Cairo, supposedly dug about 1600 BC. The first 49.5 m of the shaft comprised a spiral staircase around the shaft up which slaves carried pots of water. The lower 39 m to the water table were worked by mules turning an endless belt of leather buckets. Before the application of oil-drilling and wind-power technology, hand-dug wells through the soft loess of the Great Plains were up to 60 m deep in the 1860s–1880s, with hand windlasses to bring up the water in buckets. Even if the Romans had known of the aquifers under the Sahara at depths of 1000–2000 m, and even if they had been able to dig down to them, there was no technology available to bring that water to the surface in useful quan-tities, unless of course the water was artesian.

The great attraction of the artesian waters was not only the availability of the water at depth but that it delivered itself by artesian pressure to the surface. No wonder the geysers of the artesian bores in western Queensland (Australia), spouting over 4.5 million litres of water a day (at temperatures up to 98 °C) several metres into the air, were considered one of the wonders of the world in the 1880s!

The development of pumping technology, particularly the ability to raise water from great depth, has expanded the water resources of the arid lands. The piston valve used in most windmill pumps raises water comfortably 100–200 m and the submersible pump taps deeper aquifers, but even now

65

the mechanics of pumping are such that it is impossible to lift water efficiently from depths of over 1000 m. At such depths artesian pressure is vital if the water is to be made available.

Quality

All water contains some salts in addition to the hydrogen and oxygen ions and thus chemical reaction is inevitable in any use of water. Within the arid lands, management of water quality is obviously important to domestic users, but because of the larger quantities required the problems of quality control are greater for agricultural and industrial users. In the case of agricultural use as irrigation, the problems relate to soil:water chemistry and plant tolerances. In the case of industrial uses the concern is for corrosion problems and the demands of competing users.

Irrigation and the salinity problem

The basic problem with irrigation is that certain techniques, particularly flooding from open channels, leave water standing on the soil for several hours. This leads not only to puddling of the soil but also loss from evaporation and concentration of salts at the soil surface.

With very pure irrigation water of say 100 ppm salts, evaporation of 30 cm of water over 1 ha would leave 313 kg of salts on that hectare of land. With water slightly purer than normal urban supplies, say 500 ppm, evaporation of 1 m would leave 5500 kg of salt per hectare. In this latter situation at least a quarter of the water applied would have to be used to leach away the accumulated salts. At larger salt concentrations even larger amounts of water would be required to flush out harmful salts (Table 5.4).

Table 5.4 Control of salinization in irrigated land

Salt concentration (excluding sodium) in irrigation water (g litre^{-1})	Frequency of soil leachings needed	Drainage water removed as % of total intake
0.5–1.0	Once in 1–2 years	10–15
1.0–2.0	1–2 per year	20–25
2.0–3.0	Several times per year	30–35
4.0–5.0	Each watering	50–60

Source: White 1978.

As an illustration of the problem, in the drought of 1967–8 the River Murray contained 600 ppm salts in South Australia, and was being used for irrigation of orchards and vineyards. In 1926 during another drought the river water (already in use for irrigation) reached 1000 ppm before it even entered South Australia. With increasing demands for irrigation and domestic water for Adelaide and the river towns, the South Australian Government is examining options for salinity control whose costs range from $10 million to $200 million.

In some cases there was a dramatic chemical reaction when irrigation water was applied to land for the first time. In the initial excitement after the discovery of the Great Artesian Basin in eastern Australia in 1879, the logical hope was that irrigation would be possible in the arid continental interior. In 1894 the New South Wales Government set up an experimental irrigation farm at Pera Bore in the northwest of the State to test the possibilities. After 2 years the experiment was abandoned. The water, highly alkaline, had reacted chemically with the clay soils to produce a crust so hard it had to be broken up by dynamite to allow seeds to be planted! The future use of the water was to be for livestock rather than for crops.

Similar types of problems affect calcareous soils which are common in the arid lands. Thus, in areas of Syria, Lebanon, Jordan, northern Iraq, and Egypt (outside the Nile alluvium):

These soils are characterized by low water-holding capacity, deterioration of structure under irrigation, formation of surface crust when irrigated, and specific [adverse] hydrodynamic properties. The irrigation efficiency is low, and loss of water to the groundwater creates problems of waterlogging and secondary salinization (Elgabaly 1977: 37)

In these cases the combination of poor-quality water reacting with specific soil types can create inefficient use of the resource. The time for deterioration to become evident is not as spectacularly short as for the Pera Bore, but the suggestion is that within 14 years the salts deposited in the soil will have reached toxic proportions for plants.

Competing users of water

Irrespective of the supply, the demand for water varies considerably among the various activities of mankind. Animals, including man, may need 10 tonnes of water for every tonne of body tissue per year. Adult human daily needs vary from a European's 3.7 litres, to from 7.5 litres to about 15 litres for a Saharan or Arabian labourer (White 1960). This variation in human demand is overshadowed by the enormous variation in demands for different industrial and agricultural products (Table 5.5). The large amounts required, particularly in agriculture, emphasize not only the vulnerability of agriculture in arid lands but also reflect the enormous inefficiencies of water use here. Not only is much of the water lost from the production process by natural wastage, but much is used to maintain the environmental conditions suitable for production or to maintain the organisms which provide the end product.

Different users both require and can afford to pay for different qualities of water. Estimates of water quality required in the 1960s show the relationship between water quality required and the price which could be paid

Table 5.5 Water quantity requirements

Demand	Water needed (*tons per unit produced*)
Industrial	
1 brick	1–2
1 ton paper	250
1 ton nitrogen	600
Agricultural	
1 ton irrigated sugar cane	1 000
1 ton wheat	1 500
1 ton rice	4 000
1 ton cotton	10 000
1 ton scoured wool	>3 600 000
Animals/men	
1 ton tissue per year	1

Source: Revelle 1971.

(Table 5.6). So far, the bulk of purified water has been used for domestic or industrial purposes which alone have been able to bear the high costs. For agriculture the water demand is so high that the cost of purified water would absorb any profits from the crops. Thus a field of grain needs 2273 m^3 ha^{-1} y^{-1}. At a cost of $US1.25 per 1000 gallons and assuming 100 per cent efficiency in irrigation watering (which is never achieved) the crop would cost $250 per ha just from watering alone. On experimental plots yields of wheat have reached a maximum of 9000 kg ha^{-1}; the highest field yields have been just over half that (*c.* 4810 kg ha^{-1} in Denmark) (Watt 1973). At late 1970s wheat prices, about 9 cents per kg, these yields would give a return of $810 per ha for the experimental farm but only $432.90 per ha for maximum practical field yields. At current costs and prices, irrigation of wheat would pay only on the experimental farm. Further, with rain-fed wheat yields only *c.* 930 kg ha^{-1}, income would be only $83.70 per ha, one-

Table 5.6 Water quality requirements

User	Water quality needed (*ppm*)	Bearable cost c. 1963 (*$A per 000 gallons*)	Quantity needed
Domestic[a]	500–1000	1.00–1.50	Min.
Industrial	0–500	0.70–1.00	↓
Agricultural	200–400	10 cents	Max.

[a] Adelaide (South Australia) city water price 1976 = 72 cents per 1000 galls; 1980 = 86 cents per 1000 galls.
Source: Australian Water Resources Council (1966) *A Survey of Water Desalination Methods and their Relevance to Australia*, Hydrological Series No. 1, Canberra.

third the cost of watering. Yields would need to be tripled to pay for the water alone.

Increasing demand for water is a world-wide phenomenon associated with rising living standards and greater domestic use of water. In Iran in the 1970s average rural domestic water use was less than 50 litres head^{-1} day^{-1}. Increased standards of living in the rural areas will raise consumption to nearer the urban levels of 500 litres day^{-1} in an arid country where already water use is surprisingly inefficient (Fig. 5.1). Some 84 per cent of the precipitation is lost by evapotranspiration, some 60 per cent in fact lost from economically useless vegetation or uncultivated land. The water available for

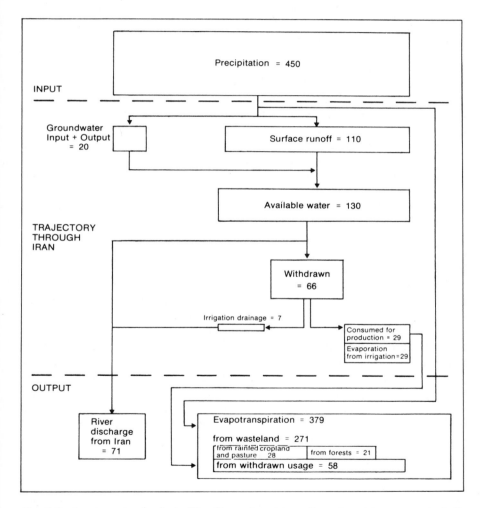

Fig. 5.1 Iran's water budget. The flow of moisture from input to output is indicated in billion m^3/yr. *Source*: Beaumont 1974

use represents 28 per cent of the input, but 55 per cent of this flows out of the country as river discharge. Agriculture uses 99 per cent of the water used by man in Iran, but the water lost by evaporation during irrigation of the fields is estimated to be 44 per cent of the total water used by agriculture and industry and 22 per cent of the total water available for human use. Obviously, there is room for more efficient use of water here.

Desalination

One strategy in water management with a long history is desalination. Various techniques have been developed to refine brackish or sea-water into potable supplies. Some, such as distillation are relatively simple and very old – being used for example by British ships transporting African slaves to the Americas in the late sixteenth century. Others such as multiple-flash or electrodialysis are complex and require large scale capital investment. All methods face the problem of cost.

Desalination plants exist around the world, not merely in arid lands, but wherever the demand for water cannot be met by conventional supplies and the users can afford the cost. Thus, most ocean liners have a sea-water distillation plant using heat from the ship's boilers to augment their freshwater tanks. Remote locations with inadequate freshwater supplies are often forced to rely upon treatment of brackish alternatives. Copper mines in the Atacama Desert were forced to distil drinking water for their workers in the 1890s (Bowman 1924). The port of Dampier in Western Australia, expanding rapidly in the iron-ore export boom in the 1960s–1970s, obtained 2.7 million litres of distilled sea-water for its new water needs by using the surplus heat from the iron ore pelletizing/roasting plants. For such remote locations there was little choice.

The main problem of water purification systems is that the costs are still higher, even in most parts of the arid lands, than that of providing supplies from conventional sources. The purification costs depend upon the technique used, the quantity and quality of water required, and the quality of the water to be treated. Estimates in 1963 gave forecasts of declining costs of water purification and parallel increases in conventional water resources so that by 1968 these costs were to be approximately equal. In fact this meeting of the costs did not take place because conventional costs did not rise, nor did purification costs decline, as fast as predicted. In 1963 desalination of sea-water had come down from $US4 to $US1.25 per 1000 gallons (0.27 cents m^{-3}) and was forecast to be $c.$ 80 cents per 1000 gallons (0.17 cents m^{-3}) by 1968 (Thorne 1963). However, in 1973 the cost was still $1 per 1000 gallons (0.21 cents m^{-3}), only half the cost in 1960 but still higher than conventional supplies. From 1973 the situation worsened as the increased oil prices forced up the costs of sea-water distillation. Estimates of these increases were 200 per cent between 1973 and 1975 and a further 15 per cent from 1975 to 1977 (Evans 1979).

Table 5.7 Desalination plants 1977–81

Area	1977			1978–81 expansion[a]		
	Plants	Capacity (000 m³ day⁻¹)	%	Plants	Capacity (000 m³ day⁻¹)	%
Middle East:	329	1822	49.2	17	1318	75.3
– Iran				1	197	
– Kuwait				2	189	
– Qatar				2	134	
– Saudi Arabia				7	539	
– United Arab Emirates				5	259	
North America	681	645	13.6	1	360	20.6
Asia	114	318	8.5	?		
Australia	7	5	0.1	0		
Africa	132	319	8.6	1	48	2.7
South America	25	28	0.8	1	9	0.5
Europe	266	450	12.1	1	15	0.9
USSR	7	115	3.2	?		
Pacific	11	6	0.1	?		
Total	1572	3708	100%		1750	100%

[a] Major installations forecast for period.
Source: Evans 1979.

By the late 1970s the location of desalination plants showed the dominance of the Middle Eastern countries, with 49 per cent of the distilled capacity in 1977 and 75 per cent of the major new capacity planned for 1978–81 (Table 5.7). Yet the global production capacity of 1977, some 2.6 million m³ day⁻¹, was not as great as had been hoped for and had not realized the forecasts of the 1950s. One recent commentator (Evans 1979: 133) suggested:

> Current state-of-the-art plants with their attendant power units are technologically complex, very capital and energy intensive, generally remain beyond the resources of developing nations in arid zones, and reflect a failure to realise the goal of cheap water from the sea.

The problems were not only related to the rise in energy costs, but also to recognition of the limitations of some previously developed technologies.

By the late 1970s, except for small solar stills whose capacity is generally limited to 20–30 m³ day⁻¹ large capacity plants relied upon three major techniques. Distillation using fossil fuels accounted for 77.5 per cent of all desalination capacity and provided 99.4 per cent of all sea-water treatments. Membrane techniques (by which water was forced through membranes to separate out the salts) provided 22.5 per cent of all the remaining capacity,

Table 5.8 Desalination costs *c.* 1977

Process	Salinity Range (000 mg litre^{-1})	Energy required (kWh m^{-3})	Capital costs ($US m^{-3} day^{-1})	Operating costs ($US m^{-3} day^{-1})
1. Distillation				
– direct (sea-water)	10–50	13.8–18.5 (thermal)	1125	1.2
– vapour compression				
(sea-water)	10–50	16.6 (mechanical)	?	1.0
2. Membrane				
– reverse osmosis				
(brackish)	0.7–10	2.1 (mechanical)	225	0.33
(sea-water)	35	5.8–8.9 (mechanical)	1030	1.06
– electrodialysis				
(brackish)	1–5	2.6 (electrical)	200	0.25
3. Freezing				
(sea-water)	10–50	12.4 (mechanical)	~800[a]	~0.7[a]

[a] Estimates.
Source: Evans 1979.

and freezing processes provided only 0.02 per cent (Evans 1979).

Between desalination of sea-water and brackish water, costs differed. The cheaper costs of brackish-water treatment using membranes have meant that the use of this process increased rapidly during the 1970s and membrane capacity by 1977 almost equalled that of the sea-water distillation plants of 1970. The reason lies in the greater energy efficiencies of the membrane processes (Table 5.8). Provided a source of brackish water is available, membrane techniques currently offer the cheapest rate of improvement of water quality.

The Israeli water budget: a case study in resource planning

One country which has applied contemporary technology most extensively and, in general, most successfully to the efficient use of its available water resources is Israel. It offers a case history (Wiener 1977) where the importance of water management required a national approach.

Approximately 76 per cent of Israel is arid or semi-arid, 61 per cent being arid or extremely arid. Rainfall reaches 700–800 mm in the north, but is only 25 mm at Eilat on the southern edge of the Negev Desert. The expansion of population by Jewish immigration after the foundation of the State in 1948 created rapidly increasing urban water demands and attempts to intensify agriculture brought further demands from the expanding area under irrigation. In 1948, however, only about 17 per cent of the potential water resources were developed and only 17 000 ha of land were irrigated.

Initially the responses to the increasing demands were *ad hoc* local decisions to use what resources were immediately available, such as springs and ground-water sources. Although a national master plan was drawn up in 1950, actual water management was local. During the 1950s the deficiencies of this purely local approach became obvious as significant shortages appeared and regional planners began to try to transfer local surpluses to meet local deficiencies within regions of the State.

Continuing demands led to the institution of a master plan in the early 1960s, when the 'national water carrier' system of trunk aqueducts, was developed. By this plan water pumped up from the Sea of Galilee (212 m below sea level) is diverted south to the main areas of urban and agricultural demand (Fig. 5.2). A national grid of water supplies, analogous to an electricity grid, was created to provide national distribution of this fundamental resource. This distribution brought most of the available water supplies into the national planning operation.

By the 1970s, the continuing increases in demands for water were threatening to exhaust the supplies and a new stage of planning began. From this time on water-resource planning included not only control of the supply, but also of the demand. For the first time attempts were made to reduce or remove inefficient uses and substitute alternative resources for water wherever possible. Improvements in the efficiency of irrigation (by adopting drip systems and electronic control of water application) – controlled environment agriculture (glasshouse or plastic-covered crops) – and a shift from low-value to high-value crops all contributed to the new strategy.

By 1975 it was claimed that these measures were using over 95 per cent of the available conventional water resources, and had allowed the increase of the irrigated area from the 1948 figure of 17 000 ha to 200 000 ha. This was associated with a twelvefold increase in real volume of agricultural production, when total population 'had quadrupled and agricultural employment rose only by 40 per cent' (Wiener 1977: 81).

With demand increases still likely into the 1980s, the next planning strategy will be to try to tap the 'unconventional' water resources – mainly the urban sewage supplies. Municipalities already absorb about a quarter of natural water supplies and by 1990 this share is expected to increase to about a third. This reclaimed sewage water will be used in industry and agriculture and will defer yet again the need for the ultimate step which would be the massive desalination of sea-water. Alongside this further evidence of technological solutions to resource management problems, the earlier strategies of increasing the efficiency of existing water usage will be continued – the high water need but low-value crops continuing to be phased out, together

Water resources

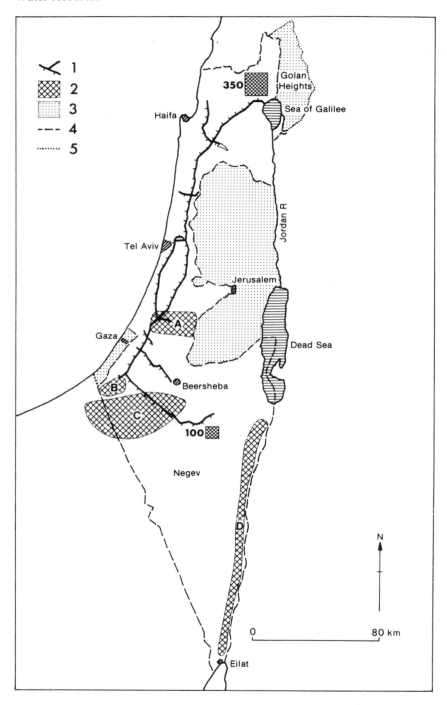

with continued improvements in the efficiency of getting water to the crops.

Israel has continued to meet the expanding national demand for water by expanding the scope of the planning process. The history of water planning shows that while the initial response was localized in space, subsequent strategies led through regional to national plans, and now the whole nation is the spatial planning unit. In time, however, expanding the spatial scope of planning for the *supply* of water was seen to be inadequate by itself, and new strategies appeared to try to control the efficiency of *use* of water. A further stage was the expansion of planning to try to reclaim and *recycle* water already in use. As yet, the next step of increasing supplies by desalination has been deferred because of the cost.

As a case history the Israeli water management story has shown:

> that water scarcity need not be a brake on economic development, nor even on the rapid expansion of irrigated agriculture, if the political will can be summoned and applied to systematic planning and scientific innovation. (Wiener 1977: 82)

The political context of the foundation and continued existence of the State of Israel has been such that national survival has depended in part upon efficient use of the national water resources. Any failure to cope here would have been equivalent to national suicide.

Fig. 5.2 Israel: national water carrier and agricultural development areas. *Source*: Amiran 1978

Key

1 – National water carrier, with intake from the Sea of Galilee indicated (350 million m^3) and volume supplied south of Beersheba (100 million m^3)

2 – Agricultural development areas: A – Lachish; B – Mirtachim (glasshouse production); C – 'Southern Project' (with recycled urban water); D – Aravah (using local springs/wells)

3 – Areas occupied by Israeli forces, mainly since 1967

4 – Border of 1967

5 – *De facto* borders of 1981. Gaza border accepted by Egypt in Camp David Accord; Golan Heights border disputed with Syria; occupied area west of Jordan River disputed with Jordan

CHAPTER 6

Plant resources

The global arid lands in the 1980s support a plant population which includes both plants peculiarly adapted to the arid conditions and, over large areas, plants which are not particularly adapted to aridity but which are maintained in the arid lands by careful soil and water management. Any planning for improved resource management in the future must try not only to improve the efficiency of this latter soil and water management, but also to understand and where possible incorporate in domesticated plants the survival strategies of those wild plants which seem to be adapted to aridity. At the same time we must realize that human occupation of portions of the arid lands goes back at least 100 000 years. Given the history of human manipulation of the environment we must expect evidence of considerable environmental impact from past as well as present human activity on the arid land ecosystems.

The discussion of the plant, and in Chapter 7, animal resources of the arid lands, will therefore consider first, the inherent adaptive survival strategies of the arid land life forms and the possible relevance for human use; second, the evidence of changes in the resources over time in the arid lands; and third, attempt an assessment of the current resources available and the possibilities for future improvement.

Characteristics of arid lands plants

Basic plant:soil:water relationships

Ecologists traditionally have recognized three basic types of plants defined in terms of their metabolism and soil water tolerances (Fig. 6.1). Within the arid lands, both mesophytes (which require moist soil conditions) and xerophytes (which can tolerate long periods of dry soil conditions) are found,

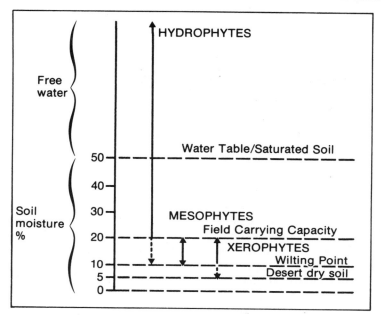

Fig. 6.1 Basic plant:soil:water relationships. *Source*: Shantz 1956

depending upon the availability of moisture from a recent precipitation or permanent ground-water.

Plant responses to soil water conditions as indicated on Fig. 6.1 can be summarized as follows. Most arid land soils contain some small amounts of moisture. Given a sufficient precipitation input the percentage of moisture in the soil rises to the *field carrying capacity* of the soil. This as a percentage varies between soils of different texture; for loam it is about 19.6 per cent, for sand it would be lower and for clay higher. At this point, if there is free drainage in the soil, further moisture will not accumulate in the soil but be lost as run-off or deep drainage. If drainage is restricted, the moisture percentage will increase until the soil becomes completely saturated (50%) and the water table is at the ground surface. Further moisture inputs will accumulate as free water on the surface.

Given this range of conditions the hydrophytes will survive in free water (e.g. water lilies) and until soil moisture drops to field carrying capacity. Below this level these plants may survive, but at the *wilting point* (10% in loam soils) the moisture in the soil is no longer available for the plants and they die. The mesophytes and xerophytes cannot tolerate the lack of oxygen in the free-water habitat nor in soil with moisture levels above field carrying capacity, i.e. where soil drainage is restricted. For the mesophytes their range of existence is possible only between 10 and 20 per cent soil moisture levels, as they also die when moisture falls below the wilting point. In contrast, the xerophytes do not die at this point, but become dormant and will only die if the soil moisture falls below 5 per cent. With renewed water availability and moisture back to 10 per cent or more the xerophytes will revive and renew their life cycle and mesophytes will germinate from seed.

Several points are worth noting here. First, no plants can survive if soil moisture drops below 5 per cent unless they have some separate water supply such as direct atmospheric moisture absorption systems. Second, the range of soil moisture conditions under which the mesophytes and xerophytes grow in any location is very small and hence small changes in moisture conditions may give spectacular results – such as the flowering of ephemeral and annual plants after a rainstorm in the arid lands. Third, if soil drainage is prevented the availability of water may not of itself be sufficient to promote plant life – it may create a local surplus of water leading to saturation, waterlogging of the soil, and death of plants by oxygen starvation.

However, ecologists have discovered that the ability of plants to cope with arid stresses reflects the system by which they take up carbon from the atmosphere. Normal humid area plants have their stomata open during the day, thereby absorbing carbon in daylight and being identified as C_3 plants. Plants tolerant of arid conditions absorb carbon at night since their stomata are often closed during the day to reduce transpiration and release that carbon more slowly to the photosynthesis process. These have been termed C_4 plants. A third category of plants, which seem to be able to adapt their carbon intake and processing to *either* the C_3 *or* C_4 system, have been identified as CAM plants (after the species *Crassulacea* – in which the Crassulacean acid metabolism (CAM) was discovered), (Adams and Willens 1978). In terms of resources, the plants able to survive the worst arid stresses seem to be the C_4 type such as salt-bushes (*Atriplex* spp.) and spinifex (*Zygochloa* spp.) or some of the CAM plants such as the *Weltwitschia mirabilis* of the Namib Desert. As a result, botanical and biological research is investigating the possibilities for breeding to transfer some of these characteristics to domestic crops.

Plant stresses in arid lands

Apart from the basic constraints of soil water conditions, the arid environment poses distinctive stresses for plant life, such as the variability of available moisture, the high insolation, threat of fire, and the effects of erosion processes (Pl. 6.1). In addition, the pressures of grazing animals, both wild and domesticated, and human harvesting pressures common to most ecosystems are equally relevant here.

The variable inputs of precipitation mean that plants must be able to cope with long periods when water supplies are unavailable, followed by periods when often massive supplies are available but for a very short time. Plants must therefore be able to absorb moisture quickly when it is available, tolerate the frequent high salinities of such water, economize on its use when supplies are not being renewed, minimize water losses, or have life spans so short that they coincide with the availability of the moisture. Such would be the stresses facing plants entirely dependent upon precipitation falling directly on the ground commanded by their root zone.

Many plants in the arid lands, however, use water collected by surface run-off from catchments larger than that commanded by their root zone, hence the lines of bushes and trees along the watercourses and the sparse growth along the ridges and watersheds (Pl. 1.1). For some plants in the arid lands along the permanent watercourses, water stress is non-existent as they are not directly dependent upon the variable precipitation at all.

For even such plants the stresses from insolation and erosion are relevant. The massive diurnal range of ground surface temperatures noted in Chapter 3 causes physiological stresses in plant tissues which, when combined with accelerated evapotranspiration by dry wind movements over the plants, can be fatal. Winds carrying soil particles can 'sand-blast' plants close to the ground or bury them in accumulations of sand and dust. The infrequent flash floods along the surface of the normally dry watercourses can be as destructive as the gentle seepage of ground-water through the underlying gravels had been beneficial. Too much water creates just as adverse conditions as too little.

Plate 6.1 Natural fire scars, Bonney Well, central Australia, photographed 1971. A series of V-shaped scars from several different lightning-induced 'bushfires' are evident. Earlier fires burned from the top left towards the camera, later fires burned from both bottom left and in one case from centre right. Despite the fact that most of the scars were over 12 months old when photographed little regeneration appeared to have taken place. Photographed from about 4000 m.

79

Plant adaptations to aridity

That such a variety of plants survives in the arid lands is testimony both to the variety of survival mechanisms in those plants and their ability to exploit the modifications of aridity over time and space.

General survival mechanisms

The various ways in which plants appear to cope with aridity include behavioural, morphological, and physiological mechanisms. Each plant uses some but not all of those mechanisms to adapt to the environmental stresses. The mechanisms include maximization of the search for water, occupation of non-arid micro-climate environments, minimization of water use and loss, tolerance of high temperatures, and carefully controlled reproductive cycles. Thus, the mesquite (*Prosopis* spp.) root systems reaching down to 80 m; the brief life spans of the ephemerals, springing up after rain but completing their reproductive cycle within a few weeks (6 weeks for the Californian Grama grass *Bouteloua aristidoides*); the waxy leaf coatings and the changing solar orientation of the leaves (e.g. some eucalypts); together with leaf drop in stress times by many plants; and delayed seed germination by the thick protective seed covers of the salt-bushes (*Atriplex* spp.) and some *Acacias* – all in their way contribute to plant survival. Summarizing these general characteristics two researchers (Solbrig and Orians 1977: 412) suggested:

> Desert plants possess special morphological characteristics, such as a low surface-to-volume ratio of leaves, reduced intercellular spaces, a relatively large root biomass . . . and increased conducting tissue, wood fibres (sclerenchyma), and green stem tissue (chlorenchyma) . . . in addition they possess special physiological traits, such as low leaf-tissue moisture and high osmotic pressures.

For resource assessment purposes, however, some classification of these methods might help managers see to what extent the plants might be available for human use.

Special survival mechanisms

In 1927 H. L. Shantz suggested a fourfold classification of plant adaptations to aridity, which is still of value to any assessment of arid land plant resources (Table 6.1). His classification distinguished some plants as *drought escapers*, growing only in non-arid conditions and surviving the long arid periods as seeds. These would include the ephemeral plants and some annual grasses and herbs whose high rates of photosynthesis provided the energy, when water was available, for rapid completion of their life cycles.

The *drought evaders* included perennials, mainly trees or shrubs, which had access through their extensive deep root systems to permanent ground-

Table 6.1 Characteristics of arid land plants

Characteristics	Plant types			
	Drought escapers	Drought evaders	Drought resisters	Drought endurers
Life span	Ephemerals; Annuals (coincides with mesophytic conditions)	Perennials	Perennials	Perennials
Phenotypic characteristic	No specialization	Phreatophytes with deep roots	Succulents; no leaves	Evergreen shrubs; small specialized leaves or no leaves
Photosynthesis	High rate when water available	No obvious specialization	Very low but possible at all times	Low but can photosynthesize under water stress
Water economy	No specialization	Tap deep underground water	Store water	Specialized to withstand water losses and stress

Source: Modified after Shantz 1956; Solbrig and Orians 1977.

water (Pl. 4.3) and thus were independent of surface soil moisture conditions. Tolerance of saline water, exudation of toxic substances to 'poison' the surface underneath their canopies for any other plants (e.g. creosotebush, *Larrea* spp.), and tolerance of considerable 'dwarfing' during water stress periods were also characteristics of this group.

The *drought resisters* were essentially the succulents and cacti which stored water in the root and stem tissues from periods of surplus for the periods of deficiency. In some cases direct absorption of atmospheric moisture via the plant tissue has been reported as for the 'rootless' cacti *Tillandiia* which appears to survive on dew-fall and coastal fog-drip in Peru.

The *drought endurers* represented those perennial shrubs which might not be tapping permanent ground-waters, but which by their very efficient control of transpiration (leaf shedding, protective coverings to leaf surfaces) and capacity to remain dormant for long periods of water deficiency remain and revive with any renewed moisture inputs.

Assessing the plant resources

The varying survival mechanisms ensure that the plant resources of the arid

81

lands vary considerably over space and through time. We need to understand these changes as background to the history of human use of those resources and before we can forecast the future potentials for that use.

Resources in space and time

In terms of available biomass over time there are obvious differences between that derived from ephemeral, annual, and perennial plants. As a result a model of responses after rains can be postulated (Fig. 6.2). While the details of the timings of the plant responses will vary among plants the basic relationships are probably valid. Certainly, pastoralists on the Austra-

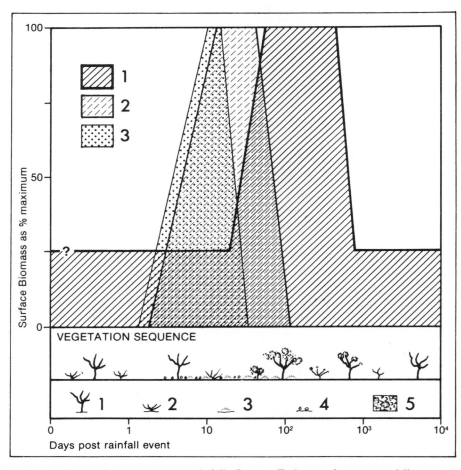

Fig. 6.2 Vegetation response to rainfall. *Source*: Estimates from general literature
Key: 1 – Tree forms, minimal leaves; 2 – Shrub forms, minimal leaves; 3 – Annual grasses and forbs; 4 – Ephemeral grasses and forbs; 5 – Significant biomass as leaves evident

lian semi-arid grazing lands consider a 50–75 mm rainfall within 48 hours will provide them with livestock feed for the next 12 months at least. The model predicts the first 'flush' of vegetation response from seeds of the ephemerals within 2–3 days, building up to a peak for these plants around 1–2 weeks after the rainstorm. Annuals might be a few days delayed in appearance, but a response from the apparently lifeless perennial woody plants (shrubs and trees) might be delayed for several weeks. With the exhaustion of the surface soil moisture and completion of their short life cycles, first the ephemerals and then the annual plants set seed, then dry up and are blown away, if they have not been eaten already. The perennials are in leaf longer since they are tapping deeper soil moisture and may offer grazing resources for 1–2 years after a substantial moisture input, before they set seed, lose their leaves, and revert to dormancy.

The spacing of plants in arid lands appears to reflect competition for available moisture. The scope for introducing further plants into the apparently unused areas therefore is limited. In an established perennial shrubland for example, while ephemerals or annuals might be found between the shrubs briefly after rains, the introduction of another perennial shrub into the space between existing shrubs will result certainly in the death of the new shrubs and possibly also severe stress on the existing shrubs because of the disturbance of their root-zone moisture catchment areas. There is also the theory that necessary nutrients are blown into the existing shrubs from these intermediate surface 'source' areas.

On a larger scale there is evidence on sand plains of the banding of vegetation into 'groves' separated by bare ground. Such groves form arcs across the slopes. Run-off from the catchments of the bare areas seems to collect and be absorbed in the groves. This alternating pattern of run-off and run-on areas, with the run-on areas carrying the perennial, often tree, vegetation has been widely reported. In Australia it has been noted among mulga (*Acacia aneura*) communities, and in Somalia and the Sahara, where the French geographers have identified it as *brousse tigrée* – tiger stripes (Pl. 6.2). Again, managers of this type of vegetation have to accept that the catchments must be retained as catchments if the tree vegetation is to survive.

Plants in the arid lands are extremely variable in the location, timing, and duration of their occurrence. To make sure that their resources are used the manager therefore must be prepared either to move rapidly over large areas as the opportunity arises to tap these, often evanescent, resources, or if limited to small areas and sedentary uses, be prepared for long periods of little or no returns.

Plants in the arid ecosystem

Such 'opportune use' may be an appropriate strategy for optimal use of arid plants, but we need to know the role of the plants in the ecosystem to assess

Plant resources

Plate 6.2 *Brousse tigrée* vegetation patterns, Bond Springs, central Australia, photographed 1971.
Alternate groves of mulga scrub (dark tone) and bare sand plain (light tone) show typical tiger-stripes (*brousse tigrée*), which are broken only by the main watercourses (darker tones) leading away from the camera; the Alice Springs to Darwin highway diagonally from lower left to right; and the Bond Springs airstrip right centre. Photographed from about 2000 m.

how much use the system can tolerate. In terms of production of dry organic matter, the arid and extreme arid lands produce about 0.1–0.5 g m^{-2} day^{-1} compared with 0.5–3.0 g m^{-2} day^{-1} for the grasslands (including the semi-arid areas) and the 10–25 g m^{-2} day^{-1} of the optimal global ecosystems, the floodplains, and estuaries (Table 6.2). As measured by net carbon production figures per unit area, the arid and semi-arid areas provide less than a fifth of the output of an average field of grain and about 2 per cent of

Table 6.2 Comparative production of organic matter in arid lands

	Gross production[a] Dry organic matter g m^{-2} day^{-1}	Net production[b] Carbon g m^{-2} yr^{-1}
Arid and extreme arid (desert)	0.1–0.5	16
Semi-arid grasslands	0.5–3.0	28
Humid grasslands		179
Grain crop	2.0	149
Estuaries/alluvial floodplains	10.0–25.0	
Tropical forest		1200

Sources: [a] Odum 1963; [b] Deevey 1971.

Table 6.3 Mineral characteristics of arid ecosystems

Ecosystem	Minerals stored in		Annual mineral movement		Nitrogen as % of total flow
	Biomass	Litter	Soil to Biomass	Biomass to litter	
	(kg ha⁻¹)		(kg ha⁻¹)		
Salt desert	143	—	85	84	13
Shrub desert	185	—	60	59	27
Dry grassland	345	70	162	161	27
Savannah grassland	978	—	319	312	26
Equatorial forest	11 081	178	2028	1540	22
Arctic tundra	159	280	38	37	53

Source: Gersmehl 1976: 229.

the output from a tropical forest. The question then arises whether these absolute figures can be changed through management of the ecosystem.

Unlike the tropical forests which contain the bulk of the nutrients of their ecosystem in the standing vegetation, the arid lands appear to concentrate their nutrients in the soil (Table 6.3).

Desert ecosystems tend to accumulate nutrients because the exits from the mineral system are blocked when moisture is lacking. The moisture deficit also hinders biological activity so the soil is the primary nutrient repository. Some desert soils are so rich that intentional leaching is necessary before crops can be grown. Extreme dryness, however, also inhibits rock weathering. The mineral cycle in a very dry environment lacks inputs as well as outputs, so an ecosystem in the core of a desert may be nutrient poor.   (Gersmehl 1976: 231)

The low biomass and scarcity of plant litter mean that although the nutrient status of the arid soils will vary considerably, these soils offer the best chance to tap nutrients from the ecosystem. However, as we shall see when the agricultural use of arid lands is discussed, the replacement of deep-rooted perennial vegetation by shallow-rooted annual grain crops has achieved increased annual organic matter production, but at the expense of deteriorating soil quality through increasing surface salinity levels.

Utilizing the plant resources

The long history of human occupation of the arid lands has meant that human use of the plants has been both spatially extensive and locally intensive. This use has, over larger areas, been destructive, as we shall see in the discussion of the environmental impact of the various systems of resource use in Part III. However, concern here is with the beneficial ways in which

Table 6.4 Plant resource uses

Main Uses	Comments	Arid land use?
(1) Food	Direct for human consumption (including plant sap)	√
	Indirect as fodder for domestic livestock	√
	Beverages	√
(2) Shelter	Construction materials for building furniture	
(3) Clothing	Fabrics, paper, silks	?
(4) Medicines	Drugs for animals and man	√
	Insecticides, fungicides	?
	Poison antidotes	√
(5) Poisons	For game	√
(6) Manures	Composts and green manures	?
(7) Industrial	Raw materials (gums, waxes, tannins, fats, alkaloids, dyes, inks, latex, creosote, cosmetics)	√
(8) Aesthetic	Ornamental species, ground cover	√
(9) Religious	Totems, sacred specimens or groves	√
(10) Environmental management	Erosion control measures ⎫ wind- Plant protection measures ⎬ breaks Cover crops, plants as buffers against environmental stress	√

Sources: Ayensu 1979 and UNESCO Arid Zone Research publications.

the plants have been used and the potentials for continued or expanded use.

The general human uses of plants are listed in Table 6.4. Most of these uses can be found in the arid lands. While the discussions of the various resource strategies will provide evidence of many of these uses, the last two decades have seen increasing interest in the arid land plants' potentials as medicinal and industrial raw materials.

Medicinal resources

Modern medicine is based upon the extensive use of therapeutic drugs derived from pharmaceutical companies, often using petroleum products and intensive capital investment in complex mechanical diagnostic devices. Currently, however, the spiralling costs of health services are being reviewed and the possibilities of cheaper services founded upon a broadly based hierarchy of simply-trained personnel trained in basic methods and with greater use of 'traditional' medicines is being considered. In this latter search the plants of the arid lands have much to offer as well as a population already experienced in using such traditional remedies.

The review of the *Medicine Plants of the Arid Zones* (Appendix 1, No. 13) pointed out the long history of plant uses for minor ailments, laxatives, and purgatives. Surprisingly, however, little of this knowledge has been followed up so far. As Ayensu (1979: 118) pointed out:

> It is astonishing and indeed an enormous human tragedy that, despite the knowledge that many modern medicines are also derived from plants, very little research has been encouraged in herbal medicine throughout the world, particularly in those regions where the health services available to the population are very inadequate

He went on to note the stimulus given to this research by the recent 'Western' exposure to traditional Chinese methods, made possible by increased professional contact from the mid-1970s.

A wide variety of therapeutic drugs can be distilled from the plants of the arid zone. Ayensu (1979: 122) suggested that:

> There are some 443 species in 64 flowering families that have medicinal value and occur in various arid regions of the world

and he provided a list of them with their proven chemical medicinal qualities, from which Table 6.5 has been abstracted. From the traditional uses to new commercial uses such as anti-oxidants, there appear to be a multitude of possibilities, with the only constraint at this stage being lack of detailed knowledge and the difficulty of distinguishing between spurious and genuine claims for the plants.

Table 6.5 Medicinal uses of some arid land plants

Plant	Location	Uses
Acacia sisalana	Mexico (Yucatan)	Disinfectant for wounds, laxative
Rauwolfia serpentia	India	75 alkaloids detected or which 3 are established tranquillizers
Commiphora mukul	Pakistan (Baluchistan)	Astringent, antiseptic, stomachic, carminative, diaphoratic, expectorant, diuretic, uterine stimulant, ulcers, bronchitis
Cassia acutifolia	Sudan, India	Laxative
Tamarindus indica	Thar Desert	Diarrhoea, dysentery, coughs, fevers, sore throat.
Larrea divaricata	Southwest USA	Rheumatism, skin disease; commercially used as anti-oxidant

Source: Plants were chosen to show the variety of uses from the list of 443 in the Appendix of Ayensu 1979.

Industrial resources

Knowledge of the potentials for industrial use of arid lands plants seems to have increased particularly after the Second World War, when shortages of strategic raw materials such as rubber stimulated interest in substitutes from the arid lands. There has been a long tradition of use of the hardwood, close-grained arid timber plants for tools, weapons, and furniture, and the fibres of the succulents and annuals for containers. Much of the research into new potentials for industrial uses has taken place in the United States, where the arid southwest and adjacent Mexican arid lands have provided both source areas and proving-grounds for many experiments.

Shortages of rubber during the Second World War, when the supplies of the *Hevea Brasiliensis* plantations in southeast Asia were denied to the Western Allies, stimulated interest in guayule (*Parthenium argentium*), a semi-arid shrub which produces rubber in the bark and leaves and which had been one of the sources of rubber for the balls in traditional Aztec and Mayan ball-court games. An Emergency Rubber Project set up in the early 1940s harvested the wild plants in the southwest USA and produced a natural rubber which proved a useful product, if slightly inferior in tensile strength. The restoration of traditional supplies after 1945, however, reduced interest in this relatively high-cost substitute. Renewed interest in the 1970s has come from fears that the rapidly expanding world demand for rubber (forecast to increase by 45% between 1977 and 1985) and the continuing high labour costs of the *Hevea* production system (some 50% from the tapping process alone) would bring shortages and unfavourable price increases. The result was an international conference on the guayule plant at Saltillo, Mexico, in 1977 and presentations at the 25th International Rubber Study Group Meeting in Washington DC in 1978. The two biggest US rubber firms, Goodyear and Firestone, both have experimental stations growing and testing the plant, and Firestone officers recently provided their forecast of the future profitability of commercial production (Table 6.6). In the words of one group

Table 6.6 Projected costs of guayule production in southwest USA 1985

Cost of factory and equipment		$35–50 million
Production costs (cents lb^{-1}):		
agricultural	29.0–38.0	
processing	20.5–28.0	
Total	49.5–66.0	
Less income from by-products		
(resins and pulp)	20.5–12.0	
Net costs	29.0–54.0	
Projected rubber price		
(1985) (cents lb^{-1})	54.0–63.0	
After tax return on an investment (%)		0–25.0

Source: Weihe *et al.* 1979: 242.

of enthusiasts (Hanson *et al.* 1979: 210):

> With a concerted effort we may expect to see a viable commercial enterprise develop around the guayule plant within 5–10 years.

Another arid plant which has come in for close scrutiny by industrialists since the 1970s is the jojoba (*Simmonsia chinensis*). Interest in the plant has brought together industrialists who have recognized the principal product, a liquid wax, as a raw material for cosmetics, pharmaceuticals, printing, and machinery lubrication. Its cultivation also has been encouraged by conservationists who see the plant's wax substituting directly for the previously vital sperm whale oil in lubrication of high heat, pressure, and oxidizing situations such as motor vehicle automatic transmission systems. From hair oil to engine oil, from cooking oil to candles, from floor polish to suncream, from protective waxes to lipsticks, from browse for livestock to ornamental garden plants, the enthusiastic supporters see few limits to its uses. Currently it is under investigation in many arid countries as a potential commercial crop in areas of 150–450 mm annual precipitation, although some researchers have claimed that optimal growth needs at least 900 mm i.e. humid conditions (Fink and Ehrler 1979)!

In some cases new commercial uses can be allied to potentials for reclamation of apparently useless arid lands. Research in Egypt has shown the capacity of several species of rushes both to colonize highly saline environments such as playas and salt-flats and at the same time to produce a useful raw material. The plants (*Juncus* spp.) withdraw salts from the soil and store them in their culms, thus 'each harvest is diminishing the salt content in soil and/or ground-water' (Boyko 1966). When treated with nitrogen–phosphorus fertilizers the plant growth was sufficient not only to reclaim the salt-flats but also to provide fibres for paper-making and in the oils from their seeds, some proteins, amino acids, and carbohydrates (Zahran *et al.* 1979).

Another area of current research is the possibility of growing phytoplankton in nutrient-enriched saline solutions. Using the techniques of the solar evaporating basins to produce commercial salt from sea-water, the method would add nutrients from treated sewage effluent to encourage growth of the phytoplankton. The result would be an algal product as raw material for vitamins, natural dyes, unsaturated fatty acids, animal and even human foods. A Mexican factory using *Spirulina* is currently producing 2 tonnes per day of dried product from 20 ha of solar ponds. In addition, some of the *Thiobacilli* oxidize insoluble metal sulphides into sulphates and allow metals such as gold, copper, iron, nickel, and uranium to be leached out of bedrock and old spoil heaps without physical mining processes. The coastal areas of the arid lands would offer the obvious sites for such solar ponds to produce these versatile but microscopic plant forms (Regan 1980).

The outlook, therefore, for industrial use of arid plants seems generally to be bright. Innovations in technology allied with changing balances

Plate 6.3 Controlled environment 'greenhouses', Environmental Research Laboratory, Tucson, Arizona, USA, photographed 1979.
The inflated plastic-covered greenhouses provide a controlled environment for a range of experiments from exploration of salt-tolerances in plants to hydroponic production of commercial vegetables such as tomatoes and cucumbers. The fans to maintain air-pressure in the greenhouses are on the left, and one of the main laboratories is on the right.

between supply and demand for traditional products and their sources have opened several avenues for possible commercial exploitation of arid plants (Pl. 6.3). The pace of scientific enquiry has increased in the last decade and commercial firms have already made speculative capital investments. The management strategies must face such questions as the efficiency of harvesting the wild plants versus plantation systems (which may have to rely upon irrigation to guarantee yields), the alternative crops to be grown on such plantations (e.g. foodstuffs versus industrial raw materials), the alternative uses of the irrigation water, and, not least, the extent to which an industrial production system can be introduced to the population living in the area.

CHAPTER 7

Animal resources

The arid lands support a sizeable proportion of the global flocks and herds (Table 7.1). To assess the efficiency of this form of resource use we need to examine the range of animal species in the arid areas, the extent of their use by mankind, and the techniques for improvement.

Animals in the arid lands

Through the work of Cloudsley-Thompson (1965, 1969), Macfarlane (1968), Macfarlane, Morris, and Howard (1963), and Schmidt-Nielsen (1965) over the last two and a half decades, we now have a reasonably comprehensive knowledge of the range of animals which inhabit the arid lands, their measures for coping with the stresses of aridity, and how they compare with domesticated animals in this regard.

There are many species surviving in the arid lands which currently have little or no value to the human population. The bulk of the insects and reptiles would fall into this category, and in some cases – the obvious one being the locusts (*Locustana* spp.) – the species comprise a definite threat to human resource use in the arid lands while the poisonous spiders, scorpions, and snakes pose local problems.

The animals which are providing resources in the arid lands are descendants of livestock which appear to have been domesticated in semi-arid to sub-humid environments in southwest Asia. Thus the asses, goats, sheep, cattle (both *Bos taurus* and *B. indicus*), and camels have not evolved from species particularly adapted to the stresses of aridity.

However, despite their non-arid origins, a significant proportion of these animals are now found in the arid lands (Table 7.1). Rough estimates of this proportion in the 1960s suggested that the bulk of the global camels, 40–50 per cent of the sheep and goats, and about a third of the horses, asses, and cattle were to be found in the arid lands. This 'invasion' had, in parts of the

Table 7.1 Livestock in arid lands

A. Estimates for 1960s

Livestock	Camels	Cattle	Asses	Goats	Horses	Sheep
Arid lands livestock (% of world total)	99	30	37	45	37	40
Proportion (%) of arid lands livestock in 'New World' arid lands (Americas, South Africa, Australia)	<1	40	<1	20	20	44

B. Livestock in arid nations (1975–9 average in 000s)

Arid nations	Camels[a]	Cattle	Mules/asses (combined)	Goats[a]	Horses	Sheep
Group I 100% arid	6 539	7 245	2 074	25 057	40	18 623
Group II 75–99% arid	6 182	113 033	11 568	106 865	3 158	309 858
Group III 50–74% arid	899	120 475	7 528	48 572	7 926	145 469
Groups I–III (% of world)	13 620 (80.9)	240 753 (19.9)	21 170 (39.3)	180 494 (40.5)	11 124 (18.0)	173 950 (45.1)
Group IV 25–49% arid	2 180 (13.0)	444 678 (36.8)	18 489 (34.3)	162 639 (36.5)	24 086 (39.0)	171 430 (16.3)
Group V 1–24% arid	247	275 181	5 761	50 349	14 484	192 968
Total arid nations (% of world)	16 047 (95.3)	960 612 (79.4)	45 420 (84.2)	393 482 (88.2)	49 694 (80.4)	838 348 (79.8)
World total	16 833	1 209 869	153 925	445 919	61 790	1 051 078

[a] 1979 figures only.
Sources: A – Heathcote and Twidale 1969; B – FAO *Year Books*.

Americas, parts of southern Africa, and Australia, been taking place only within the past 150 years and represented a major modification of the patterns of global grazing over that period.

By the 1970s the core and predominantly arid nations still held 81 per cent of the world's camels, 41–45 per cent of the goats and sheep, 39 per cent of the mules and asses, and 18–20 per cent of the horses and cattle. When the figures for the semi-arid and peripherally arid nations were added, the proportions were generally increased to 79 per cent or above. However, most of the livestock in these nations were located outside the arid zones and only a small proportion, perhaps less than a quarter, were grazing arid pastures.

An additional component to these livestock figures which is locally important but which never gets into the official statistics, is the feral animal population. Included here are wild camels in central Australia (currently some 'tens of thousands' according to McKnight 1969), wild donkeys, goats, and asses in central Australia and the southwestern USA, as well as a much less visible but ecologically significant population of wild cats and dogs. The significance of these populations lies not only in their competition with domestic livestock for the scarce feed which increases the possibility of inadvertent overgrazing, but in the case of the introduced carnivores (cats and dogs) their depredations on the indigenous mammals and modification of the balance of local ecosystems.

Domestic livestock on the arid ranges

Given that most domestic livestock in the arid lands are not particularly well adapted to the stresses of the environment, management for most efficient use of the available resources must recognize the varied capacities of the livestock to cope with those stresses – specifically of water and feed shortages.

Water stresses

Comparing the capacity of domestic livestock to survive periods without water, Schmidt-Nielsen (1956: 380) calculated the contrasts between sheep, donkeys, and cattle (Table 7.2). Noting that donkey meat is not usually eaten for religious reasons, he concluded:

> From the theoretical considerations it seems amazingly clear that the camel offers a most obvious solution to increased meat production in arid zones with a low natural vegetation density that cannot easily be increased.

Yet the camel breeders of southwest Asia in the 1980s prefer to use their oil income profits to purchase sheep from Australian semi-arid and sub-humid ranges rather than breed up their own camels for slaughter. Management involves more than the obviously rational strategies.

Where underground water can be obtained reasonably efficiently, the spacing of wells becomes an important management strategy. Most herbivores graze into the wind so that the range upwind (in terms of the *prevailing* wind) will always be more heavily grazed than that downwind – a fact confirmed by recent satellite imagery of Australian fenced ranges. Further, since livestock lose condition if they have to travel too far to water, permanent watering points should not be spaced more than 10 km apart for sheep and

Table 7.2 Grazing areas and productivity

Livestock	Maximum summer period without water (days)	Radius of range from water point (ratio with sheep range = 1)	Livestock Production from one water point	
			Number	Meat (kg)
Sheep	3 ?	1	10	250
Donkeys	4	1.3	7	630
Camels	12	4	19	8500

Note: Adequate fodder is assumed.
Source: Schmidt-Nielsen 1956: 379.

48 km apart for cattle. Having said that, however, pastoralists recognize that animals may need very little water if the feed comprises succulents (such as parakeelya, *Calandrinia* spp.) and will at such times range much further afield.

Feed stresses

As a means of converting solar energy stored in the plants, the larger herbivores used by man are relatively inefficient. The basic pyramid of numbers (of energy transfers across the trophic levels of the environment) means that less than 1 per cent of the incoming sunlight energy would ever be available to the herbivores, and in practice the proportion is much less. Love's diagram of the energy flow through the rangelands of the western USA, which includes semi-arid conditions, suggests that the energy available as meat on the live cow or steer represents only about 0.0025 per cent of the incoming solar radiation (Fig. 7.1). On better-quality grazing lands of the sub-humid to semi-arid Central Valley of California the proportion was slightly higher – 0.04 per cent of solar energy.

Efficiency of conversion of feed to useful by-products varies also among animals. The pig is generally regarded as most efficient with conversion rates of 17–20 per cent of feed into body weight; dairy cows convert 12 per cent, but steers convert only 5 per cent of their feed, while sheep are still lower at about 2 per cent. The value to man therefore of the livestock rests more on their ability to convert otherwise relatively useless vegetation into a preferred food supply than their efficiency of its conversion. Livestock grazing of natural fodder whether *browse* – twigs, leaves, and bark, high in cellulose, or *forage* – the grasses, herbs, and forbs at ground level which have much less cellulose content, make available to human use resources which would be otherwise impossible to harvest or use as a food supply except at an unbearable cost. To the extent that the livestock can convert the cellulose content (indigestible in man) into meat or milk products, they are in addition rendering an inedible item of the environment edible and therefore a resource.

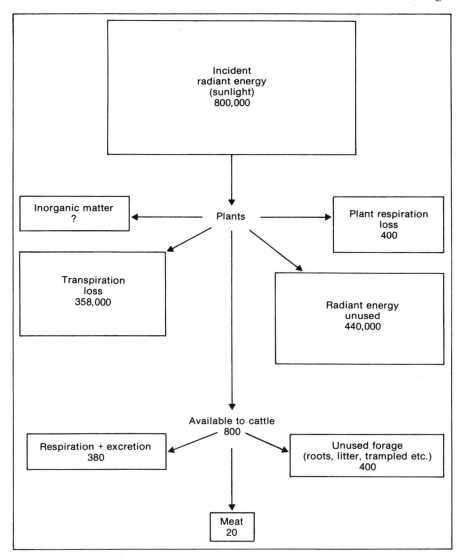

Fig. 7.1 Energy flow through the western rangelands of the USA. *Source*: Love 1971

Key: Figures as kcal m^{-2}

Note: This flow produces the equivalent of 1.56 kg of beef on the hoof per hectare

Grazing preferences

The impact of the grazing animals on the environment is related to their density and feed preferences. From domesticated livestock to wild animals there appears to be a variety of feed preferences (Table 7.3). The forage,

Table 7.3 Dietary preferences in larger herbivores

Animal	Dietary preference (% share = min. + max. figures given)		
	Grasses	*Forbs*	*Browse*
Cattle:			
Hereford	35–71	23–41	6–25
Santa Gertrudis	49–81	17–43	2–15
Sheep	37–82	1–47	16–31
Mule deer	2–6	4–42	50–94
Goats:			
Rocky Mountain	58–76	14–23	1–25
Angora	40–65	4–25	27–49

Note: Data are for USA, maximum and minimum over four seasons
Source: Heady 1975: 48–9.

grasses, and forbs, usually are preferred by the cattle and sheep while the deer prefer to browse. The effect of introducing domesticated livestock is to place enormous pressure on the ground-level vegetation. At times of drought, however, the first component of the vegetation to be affected is precisely this cover of annual or even perennial grasses, forbs, and herbs.

Studies of rumen samples from central Australian cattle in the 1960s showed that grasses dominated the diet as long as they were available and the ingestion of browse 'top-feed' (leaves of trees such as *Acacia aneura* and *A. Kempeana*) was a sign of stress in droughts which had removed the grass layer. When less than 90 per cent of the cattle feed comprised grasses, the suggestion was made 'that management should seriously consider reducing stock numbers '[to prevent drought losses] (Chippendale 1968: 27).

Some attempts to 'improve' arid range grazing capacities have involved partial processing of the fodder before it is made available to the livestock. Processing of mesquite trees – 'the entire tree chopped to sawdust-fine, hay-like feed' – as basic roughage to provide part of a diet (with grain supplements and molasses) for cattle in Texas was reported to have been successful, but the cost of such processing and the energy consumed are high (ICASALS 1968). The thinking behind such experimentation reflects in fact a narrow commitment to one type of resource use by one animal. A superior strategy in terms of energy cost would be to graze deer or goats on the mesquite range, but the traditional cattle rancher would be unlikely to agree.

The unseen herbivores

One type of grazing animal whose activities are receiving greater attention from the range managers is the termite or ant population (Lee and Wood 1971). Research in Australia, stimulated by graziers' complaints that their paddocks were remaining bare long after domestic livestock had been removed, revealed that local termite (*Drepanotermes* spp.) colonies were each harvesting 1 kg of grass per year. During a 'good' year, i.e. above average rainfall, up to 375 nests per ha could be recorded, which meant that 375 kg ha^{-1} of grasses could be 'lost' to the graziers. What the graziers lost the soil gained. Estimates put the 'turnover' of soil through the nest-building activities as 80 000 kg ha^{-1} together with the storage of the harvested grasses below the surface (Watson and Gray 1970). However, erosion was accelerated, since the nests formed a hard clay surface 2–3 m in diameter at or close to the surface and at maximum densities covered virtually the whole ground surface, when vegetation colonization would be restricted, run-off maximized and wind and water erosion encouraged. Bearing in mind the extensive and even more spectacular forms of termite nests in the dry tropics and the activities of rodents such as gophers in the western USA, it is obvious that any management of arid grazing resources must take this 'hidden' grazing pressure into consideration (Pl. 7.1).

The environmental impact of grazing

The micro-manipulation of the soil and ground surface by the termites is more than matched by the trampling of the larger animals. American research has suggested that 23 per cent of the standing crop of feed can be destroyed by trampling. A further effect is that soil is compacted (leading

Plate 7.1 Insect and animal trails in the Kara Kum Desert, USSR, photographed 1976.
Early morning evidence of the arid lands' 'night life'. There appear to be tracks of at least two different beetles, a small lizard, and two sets of small rodent prints in this area of wind-blown sand about 2 × 1½ m.

97

to reduced rainfall penetration and greater loss by run-off) by the hooves of sheep and cattle, which exert pressures equivalent to from 0.65 to 1.7 kg cm^{-2} respectively compared with up to 0.63 kg cm^{-2} for crawler tractors and 1.4–2.1 kg cm^{-2} for wheeled tractors (Heady 1975: 60). In Australia, in particular, this is a significant factor since the heaviest indigenous mammal – the kangaroo – is relatively soft-footed and only exerts 0.1 kg cm^{-2} pressure. The introduced livestock therefore had a considerably greater impact on the surface than the original herbivores.

The role of one animal – the goat – in the arid lands has been the source of controversy. On the one side the conservationists see it as the enemy of all vegetation and the cause of much of the deterioration of the arid lands. The defenders of the animal have pointed to the goats as food supply in North Africa and southwest Asia (where populations of four goats per human are not uncommon) and particularly to its ability to harvest plants from rugged mountain areas inaccessible to man:

> Nearly ominivorous, and agile to the point of overcoming all practical transportation difficulties, he extracts and concentrates the nutritive elements of an otherwise uninhabitable landscape.   (Kolars 1966: 583)

For the subsistence peasant concerned for today's food supply, the long-term

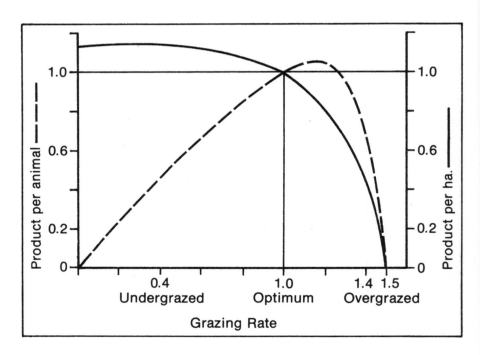

Fig. 7.2 Optimal grazing rates. *Source*: Heady 1975.
Graphs show trends of productivity per animal and per hectare related to grazing rates

impact of his goats upon the environment is irrelevant. Conservation of his resources, in the sense of deferred use, is not an option for him.

The net result of the grazing pressures may be significant environmental degradation through loss of vegetation and associated soil modifications. Heady's graph suggests the relationships between stocking rates and product of the grazing process based upon American experience (Fig. 7.2). Two examples of the apparent effects of grazing using different measurement techniques are relevant here.

South Africa
In South Africa at the time of the First World War, fears were expressed that the Kalahari Desert was expanding eastwards into the grazing and agricultural areas, particularly in times of drought. Investigations were made in 1914, again in 1920, and yet again in 1948, each as the result of the expression of popular concern at an apparent deterioration of environmental conditions. While there appeared to have been no significant change in the rainfall patterns in space or time there had been:

> a serious deterioration of the vegetation and hydrological conditions. And everywhere there has been almost terrifying erosion of the soil . . . all the damage that had been done could be explained in terms of the impact of man, with his plough and his domestic animals, upon a land that was from the very start extremely vulnerable.
>
> (Kokot 1955: 407)

The deterioration of vegetation was demonstrated by comparing past with present plant distributions, particularly noting the introduction of desert succulents in place of the grasses (Fig. 7.3).

The photographic evidence
In some of the arid lands photographic evidence allows historical comparisons in vegetation patterns to be made. Hastings and Turner (1965) duplicated early surveyors' photographs of the Arizona deserts and thus captured the changes from the period of the 1880s to 1960s. A similar study of southern New Mexico 1858–1963 showed that the mesquite trees had increased tenfold on sandy soils and the creosote-bushes twentyfold on shallow soils, both apparently as a result of the removal of competing grasses by constant grazing and periodic droughts (Harris 1966). Experiments in which grazing was excluded from reserved areas to allow natural regeneration to appear have demonstrated that regeneration is possible – for example Omdurman in the Sudan (Hills 1966) and the Tunisian Pre-Sahara Site (Bedoian 1978), but the progress can be so slow as to be irrelevant for all practical planning purposes.

A study site in South Australia showed that while some of the bushes such as *Cassia* spp. grew 2 m over the 35-year observation period, most of the trees showed hardly any growth, and this despite the exclusion of all domestic livestock for the period (Hall, Specht, and Eardley 1964). Hopes for a rapid recolonization of arid lands vegetation once grazing pressures have been removed must therefore be muted.

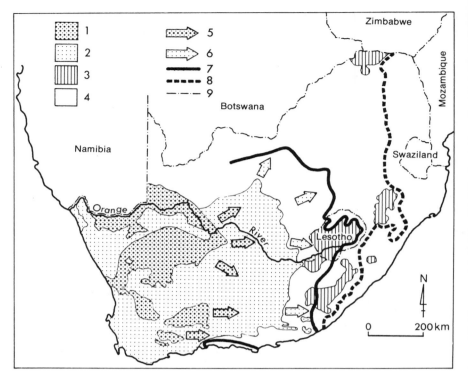

Fig. 7.3   Desertification in South Africa *c.* 1955. *Source*: Kokot 1955.
   Key: 1 – Desert; 2 – Vegetation below critical stage – maximum potential erosion; 3 – Vegetation patchily below critical stage; 4 – Mainly above critical stage; 5 – Main desert expansion; 6 – Main Karoo expansion; 7 – Limit of Karoo patches; 8 – Limit of Karoo pioneer vegetation; 9 – State boundaries

Beyond the domestic livestock

In his introduction on arid fauna as human food, Lowe (1968: 586) quoted an 1896 description of the animal foods available to central Australian aborigines that:

> To mention the names of all that are eaten would be largely to recapitulate the zoology of the district.

He suggests that such a statement would have been and, in some areas still is, true for other desert dwellers. Yet on the global scale today mankind is using as food sources only 50 of the more than 1 million species of animals (and only about 600 of the 350 000 plant species). This deliberate restriction of the choice of resources applies also to the contemporary arid lands and the researcher would have great difficulty in finding any arid area in the

100

Table 7.4 Domestic livestock and wild ungulate productivity

Species	Liveweight gain per day (kg)	Age at adult weight (months)	Average adult weight (kg)
Thomson's gazelle	0.04–06	18	18.0–24.0
Impala	0.09–12	24	45.8–59.4
Grant's gazelle	0.10–12	24	45.8–60.2
Topi	0.15–20	30	114.3–132.4
Wildebeest	0.20–24	45	163.3–208.7
Eland	0.24	48	283.5–376.5
Sheep	0.05	18	20.0–45.4
Cattle	0.14	60	158.8–453.6

Source: Watt 1973.

world in the 1980s where the plant and animal resources are being used fully and completely to support human life.

The reasons for this conscious restriction of choice seem to lie in traditional usage systems, cultural and religious dietary preferences, and taboos and ignorance of the potentials of the variety of arid animal life forms. There is no doubt that existing domestic animals could be used more widely in the arid lands than they are at present. A world map of pig distribution shows gaps in the patterns within the Muslim arid land countries of North Africa and southwest Asia. The map of goats shows similar gaps for the arid areas of the Americas and Australia. There are no real environmental barriers to such animals in these areas. The barriers are man-made as Simoons (1967) showed in his study of dietary taboos, *Eat not this Flesh*.

Of the larger animals not currently used, several have been shown to be more efficient in protein production than domestic livestock. Game animals such as gazelles, the wildebeest, and eland have been suggested as alternatives to cattle and sheep for the semi-arid country (Table 7.4).

There can be no doubt that superior livestock could be domesticated for the arid ranges. The problem is less their tolerance of aridity than human tolerance of them as resources.

CHAPTER 8

Energy in the arid lands

The energy question is a complex one. There are many diverse actors –
consuming nations who act independently of each other; producing
nations; multi-national producing companies; smaller independent
companies, diverse substitutes, etc. (Erickson and Waverman 1974: ix).

The complexities of the global energy crisis as it became identified in the
early 1970s have been reflected in economic and political activity since that
date. Much of that activity has focused on the arid nations as both major
producers and consumers of energy. We need to understand both the
reasons for that activity and the effects upon the management of energy
resources in the arid lands. We need to examine first the energy crisis, then
review the energy resources of the arid lands and their role both within the
arid lands and as part of the global energy flows.

The energy crisis

The period from the 1950s onwards saw a rapid escalation in use of fossil
fuels, particularly petroleum and its by-products, in the industrialized
nations. The use of energy on a global scale had been increasing throughout
the nineteenth century both as a result of the impact of coal-fired steam
power on manufacturing and on the establishment of world-wide sea transport based upon coal-fired steamships. The development of the internal combustion engine and its application to freight as well as passenger transport
in the early twentieth century further increased energy demands, as did rising standards of living in urban areas, the reticulation of domestic gas and
electricity supplies, and the use of the coal and petroleum fuels also as raw
materials for products ranging from cosmetics to plastics. As perhaps the
extreme case of changing energy usage, the USA data illustrate the trends
away from animal to alternative energy sources (Table 8.1).

Table 8.1 Trends in energy use in the USA, 1850–1970

Date	Population (10^6)	Total energy production[a] (10^{15} Btu)	Energy production per person (10^6 Btu)	Percentage from animals
1850	23	0.3	13	0.5
1910	92	14.0	152	0.25
1940	132	25.0	189	0.004 5
1970	203	61.0	300	0.000 08

[a] From coal, oil, gas, and water power.
Source: After Watt 1973: 142.

Much of the increase in energy consumption was of petroleum. Global consumption of fuel oil in the 1970s was increasing at a rate of 6.9 per cent per year (thus doubling itself every 10 years). Most of the consumption in the industrial nations was in three sectors of the economy: manufacturing, electricity generation, and transport. The detailed consumption pattern for

Table 8.2 Energy usage in Australia in the mid-1970s

A. Usage by sector of economy

Sector of economy	Percentage of national energy
Manufacturing	33
Electricity generation	28
Transport and storage	27
Domestic and commercial	5
Mining	3
Agriculture, forestry, fisheries	2
Other uses	2
Total	100%

B. Source and usage of energy supplies

Supply	Percentage	Usage	Percentage
Oil	48	Transport	40
Coal	40	Heating	
Natural gas	7	(industrial	
Miscellaneous	5	and	
		domestic)	50
		Electricity	10
Total	100%		100%
Amount supplied	2.8×10^{18} J yr^{-1}	Amount used[a]	2.3×10^{18} J yr^{-1}

[a] A loss of 0.5×10^{18} J yr^{-1} is recorded.
Sources: A. *Senate Report on Solar Energy*, 1977; 5; B. Diesendorf 1979: 27 and Morse 1977: 210.

Australia illustrates the relatively small proportions going into agriculture, mining, and domestic uses (Table 8.2). Table 8.2 also illustrates some of the complexities of energy use accounting, particularly where fossil fuels are transformed into electricity to be used for a variety of purposes from transport to heating.

The energy crisis came not from a recognition of a shortage of supplies as a result of this increasing demand, but from a political decision by the Oil Producing and Exporting Countries (OPEC) to increase the price of crude petroleum in 1973 and to forecast future price rises, as a means to increase their national revenues. The industrial nations having built up their manufacturing expansion on the basis of cheap oil were caught out, the cost of their main raw material and energy supplies went up overnight, and they began to search out and reassess alternatives. This frantic search to reduce the inflationary effects of a massive increase in oil prices was the crisis for the industrialized nations. For the oil producers, 1973 initiated a bonanza period of expanding national increases and political influence. As we shall see several of the arid nations directly benefited.

Energy resources in the arid lands

One of the effects of the energy crisis was to crystallize thinking on the alternative sources of, and uses for, energy on a global scale. The main concern of such thinking was to make most efficient use of the different sources, each of which had their optimal application.

Three broad sources of global energy were recognized: gravitation, nuclear, and solar, and for each optimal uses might be indicated (Table 8.3). In terms of duration, the gravitational and solar flux sources offered long-term supplies while the geological storage of the solar sources as fossil fuels and the nuclear minerals in the earth's crust offered relatively short-term supplies. How do the resources of the arid lands fit into this global picture?

Gravitational energy

In the monograph on *Coastal Deserts* (Amiran and Wilson 1973), there is no mention of the potentials for use of tidal power as an energy source. Yet the potential of the tides exists and may be locally important in the future.

On the global scale only one tidal-powered generating station exists, at La Rance on the west coast of France. Established in 1966 it produces up to 240 MW from a tidal range of 8.4 m with an efficiency of 18 per cent (Woods 1977). There appears to have been serious study of the tidal potentials only on the arid coasts of northwestern Australia where a tidal range of 12 m at Secure Bay (Western Australia) could support a power station producing up to 520 MW. In 1976, the construction costs of such a station (some $400 million) would have been three times that of a conventional coal-fired electricity generating station.

Table 5.5 Global energy sources and applications

Energy			Optional energy application				
Source	Expression	Duration[a]	Mechanical	Electrical	Thermal	Chemical	Material
1. *Gravitation*	Ocean tides	P	✓				
2. *Nuclear*							
– earth core	Geothermal:						
	– steam	T			✓		
	– hot rock	P			✓		
– earth crust	Fission:						
	– burner	T			✓		
	– breeder	T			✓		
	Fusion:						
	– lithium–deuterium	T			✓		
– ocean	– deuterium/deuterium	P?			✓		
3. *Solar*							
– flux	– Wind	P	✓				
	– Ocean current	P	✓		✓		
	– Hydro	P	✓				
	– Wave	P	✓				
	– Thermal gradient	P			✓		
	– Radiant:						
	• photovoltaic	P		✓	✓		
	• photothermal	P		✓	✓		
	– Biological storage:						
	• photochemical (food, fibre, minerals)	P				✓	
	– Wastes	P			?		
– geological storage	– Natural gas	T				✓	
	– Oil	T				✓	
	– Tar sand	T				✓	
	– Oil shale	T				✓	
	– Coal	T				✓	

a P – Permanent supply; T – Temporary or diminishing supply.
Source: Australian Conservation Foundation, *Tjurkulpa*, **11**(8), 1979, 3.

The tidal generating stations then are technically possible. However their operating efficiencies are low (because of the periodic nature of the supply), and their high capital cost, limited potential sites, and relative cheapness (so far) of alternatives will mean that they remain on the drawing-boards for many years.

Nuclear energy

In the mid-1970s approximately 57 per cent of the world's uranium production was mined in the arid lands, with the USA providing 44 per cent, South Africa 11 per cent, and Niger 7 per cent. Canada also provided 25 per cent, but this was from mines outside the arid zone. Reserves of uranium in the arid lands were about 76 per cent of world figures, with the USA holding 30 per cent, Australia 22 per cent, South Africa 17 per cent, and the North African arid nations 6 per cent. Little direct use is made of uranium for energy supplies within the arid lands. The bulk of the production is exported for use in the more humid areas.

Solar energy: fluxes

Two basic types of solar energy can be distinguished: a variety of energy flows (fluxes) which can be directly measured and used as sources, and the stored solar energy trapped in the deposits of coal, oil, and natural gas (Table 8.3).

Wind power
While the details of performance and reliability of wind-driven machines often are disputed, there is no doubt that wind power is at work in the arid lands in the 1980s and its scope could be widened.

As noted in Chapter 3, the arid lands are not especially endowed with continuous strong wind movements. Nevertheless, the traveller in the arid lands will be impressed by the use of small wind-pumps for raising water, the ancient wind-driven grain grinding mills of Iran and Pakistan and traditional architectural styles designed to take advantage of prevailing winds for cooling and ventilation as in Egypt and southwest Asia (Cain *et al.* 1976). In the USA, since the mid-nineteenth century over 6 million windmills of less than 1 hp have been built and 150 000 are still in operation, most raising underground water to the surface and many to be found in the semi-arid and arid lands of the nation. A 24 km h^{-1} wind acting on their 3.6 m diameter blades provides about 0.16 hp, sufficient to raise 159 litres of water per minute to a height of 8.7 m (Eldridge 1976).

Research in the early years of the UNESCO Arid Land Research Programme suggested that a 15 m diameter rotor working in an average 20 km h^{-1} wind could provide 104 000 kWh yr^{-1}, which would be sufficient for lighting, water heating, pumping, and refrigeration for a village of 100 families (Golding 1953: 599). Research has swung towards the larger plants, e.g. the Tvindmill in Denmark with 54 m diameter blades on a tower 53 m high. Wind speeds of 15 m sec^{-1} provide 2 MW, and costs of electricity generation were given as $440–480 per rated kW compared with electricity from a nuclear power plant at $1000 per kW (Diesendorf 1979).

That wind power is not more widely used on the arid lands seems to be the result of the cheapness (until the 1970s) of alternatives (whether wood or petroleum) and problems of storing energy in batteries associated with the irregular occurrence of winds and their relatively low velocity. The suggestion has been made that the engineers were about to make a major breakthrough in design of wind machines in the 1950s when, almost overnight, the price of oil fell as a result of major discoveries in the arid countries of the Persian Gulf. With cheap oil available, the incentive for wind-power research was removed (Eldridge 1976). It remains to be seen whether the increase of oil prices in the 1970s will revive hopes of that breakthrough. Until that may happen, however, wind power will continue to be the small-scale alternative to conventional fossil fuel energy sources, particularly in the remote and isolated areas of the global arid lands.

Solar radiation
The characteristics of direct solar radiation are the low energy input per unit area, averaging 0.15–0.20 kW m^{-2} h^{-1} with a maximum of 1.12 kW m^{-2} h^{-1}; the intermittent and often periodic inputs, between daylight and night-time periods, and high absolute inputs. For example, the Atacama Desert was estimated to receive in one year the equivalent of all the fossil fuels used in the world in the mid-1960s (Daniels 1967). Use of such radiant energy faces problems of concentration of the supply, storage for the non-productive periods, and varying efficiency in usage. The latter ranges from efficiencies (percentage of energy received versus energy used) of 0.025 per cent for plant photosynthesis, 10 per cent for silicon photovoltaic cells, 25 per cent for solar refrigeration, and 40 per cent for water distillation. In the mid-1960s, 9000 m^3 of water per year were produced by solar distillation and locally such plants have a vital role, as at the opal mining town of Coober Pedy in South Australia where a distillation plant forms the main water supply.

Various proposals have been made to produce electricity by massive arrays of photovoltaic cells or reflectors focusing the sun's rays on to 'power towers' holding the cells. In the USA two such towers already exist, one of 5 MW output at Albuquerque costing $21 million and one of 10 MW in California costing $120 million. As yet they are still in the experimental stage. Other experiments are concerned with attempts to store and transport the energy in a chemical form, as hydrogen or ammonia. Pilot plants for such processes were currently planned to open in Australia in 1982 (*The Australian*, Sydney, 26 March 1977).

Actual use of solar radiation in the arid lands is extensive and is likely

107

to increase in the future. Sun-dried fruits and solar evaporator pans for salt are traditional uses still operating. Solar hot-water heaters are becoming increasingly popular as electricity costs increase, especially where the electricity is generated by oil-fuelled power stations, and cheap and simple solar cookers have been developed in India. Direct generation of electricity from photovoltaic cells is in limited commercial use already. In Australia remote microwave-VHF repeating station links in radio communications systems are already using solar power to maintain batteries and provide basic power needs. Experimental solar-powered air-conditioned houses have been built and a combined wind and solar-powered house is currently under test in South Australia. Solar-powered water pumps were installed in Abu Dhabi in 1978 (Cordes and Scholz 1980: 41). On the Papago Indian Reservation west of Tucson, Arizona, a solar-powered village, claimed to be the first in the world, was opened in 1980. Here combined photovoltaic panel and storage batteries system provided up to 17 kWh day^{-1} for lighting, water pumping, and refrigeration (*Arid Lands Newsletter*, 1980: 28).

The cost of the direct production of electricity from solar radiation has been dictated by the high cost of the silicon-coated photovoltaic cells. If this price could be reduced along with reduction in the cost of storage batteries, then we might forecast a significant increase in use of solar electricity, not least to reduce the demands upon coal. Whereas 1 kWh of electricity generated from coal needs 3 kWh of coal, if that same amount of coal were used to produce solar heaters some 15 kWh of electricity would result (Gartside 1977).

Biological storage of solar power

The energy crisis also stimulated interest in the conversion of plants and plant residues into energy supplies. Traditionally in the arid lands, woody plants together with dried animal dung have been the main source for domestic fuel supplies. Father Carpini, Venetian envoy to Ghenghis Khan in AD 1162, noted that from the Emperor down, all cooking and space heating was by fires of horse and cattle dung (Lamb 1963). For a brief period on the Great Plains in the mid-nineteenth century 'prairie coal' – dried buffalo or cattle dung – was widely used. Locally the cutting of firewood played a more important role, not only as a fuel source but in terms of its impact on the local vegetation. Estimates for western North Africa put the individual daily needs of wood at 1 kg, which represents an average of 0.5–1 ha of woody vegetation destroyed per person per year (Le Houerou 1970). In the Sudan figures show that *per capita* consumption of wood for fuel has increased from 1.62 m^3 in 1962 to an average of 2.0 m^3 in 1976–7, with consumption in the arid areas up to 2.8 m^3. In fact, consumption has long since outstripped the local supply and of the 11 million m^3 national wood 'budget' consumption in 1976–7 some 4 million m^3 had to be imported (El Arifi 1979). The advent of cheap kerosene in the 1950s reduced this pressure on the environment somewhat, but the increased prices of the 1970s have brought further cutting and the fears of desertification.

Use of plants or plant residues to make alcohol came under renewed scrutiny in the 1970s. Most studies have been oriented to cropping in the humid lands, although some have suggested using part of the straw from the semi-arid wheat fields as a basic source of methanol. Serious proposals have been made in countries such as Australia that the existing food grain area should be extended to produce grains for conversion to ethanol. Other possibilities are the cultivation of plants which contain hydrocarbons as a milky sap, or latex. Such plants include Euphorbiaceae (e.g. milk-weed) and Asclepiaeceae (e.g. cotton bush) which can tolerate arid conditions.

Most of the arguments are based upon inadequate data, but there are pilot processes already producing alcohols from sugar-cane residues and cassava so the potential for transfer of that technology to arid lands exists (CSIRO 1979). However, the sparse yields per unit area and the absence of major demands for energy *in* the arid lands are not likely to encourage the sowing of crops for fuel supplies. Suggestions for removing the wheat stubbles from the semi-arid fields already have brought strong protests from soil conservationists who point out the important role of those stubbles as protection against wind and water erosion.

Wastes

As noted above the use of dried animal dung as a domestic fuel is widespread in the arid lands of Africa and Asia and was used briefly in North America. Dried human excrement is used to heat the public baths of Sana, Yemen (Kirkman 1976), but most recent research has concentrated upon indirect use to produce methane gas for domestic cooking.

Fossil fuels: geological storage of solar power

Three fossil fuels derived from geological storage of solar energy can be identified in the arid lands (Table 8.4). The oldest in terms of length of use is crude oil. There is evidence of the use of oil seeps in Mesopotamia as early as 3000 BC, but the effective widespread use of oil and natural gas in the arid lands did not come until after the Second World War.

Coal
Coal production in the predominantly arid lands in the 1970s represented only *c.* 7 per cent of the global production. It was the production from the semi-arid nations, mainly China and the USA, and the peripherally arid nations, dominated by the USSR, which provided the other 70 per cent of the global total. In the core arid nations no coal production at all was recorded, the energy supplies coming from oil or natural gas.

The major arid nation coal producers in 1977 were as follows (in million

Table 8.4 Fossil fuel energy in arid lands

| | Fossil fuel energy sources[a] | | | | | |
| | Petroleum | | Natural gas 1977 | | Coal | |
Arid nations	Reserves 1977 10^6 tonnes	Production (Average 1974–7) 10^6 tonnes	Reserves 10^9 tonnes	Production (tetracalories) 10^3	Reserves[b] c. 1970 10^6 tonnes	Production 1977 10^6 tonnes
Group I 100% arid	30 642	671	5 810	208	13	0
Group II 75–99% arid	16 754	570	16 918	481	14 901	74
Group III 50–74% arid	326	24	240	65	10 818	91
Groups I–III (% of world)	47 722 (64.0)	1264 (44.8)	22 968 (36.4)	753 (6.4)	25 732 (6.0)	163 (6.7)
Group IV 25–49% arid	8 488	217	7 722	5 142	250 275	1202
Group V 1–24% arid	13 446	824	26 359	3 712	91 537	531
Total arid nations (% of world)	69 656 (93.4)	2305 (81.6)	57 050 (90.4)	9 607 (81.3)	367 544 (85.5)	1898 (76.6)
World	74 587	2824	63 108	11 820	430 101	2476

[a] Figures are rounded and in some cases known producer nations were omitted from source data.
[b] Recoverable reserves.
Sources: Statesman's Year Books and UN Statistical Year Books.

tonnes): the USSR 499, South Africa 85, Australia 71, the USA 60, Canada 23, and India 10. However, the bulk of coal production comes from outside the arid areas of each nation. Even where production comes from within the arid lands it is usually exported for use in other humid portions of the nation. Thus the opencast coal mines of the Great Plains of the USA supply the eastern metropolitan power stations and the steel mills of Pennsylvania; the South Australian Leigh Creek sub-bituminous coal supplies the Port Augusta electric power stations which in turn feed into the State's grid system serving the humid fifth of the State.

In terms of recoverable coal reserves the predominantly arid nations have the same global proportion (6%) as their production, but the semi-arid and peripherally arid nations increase the total arid nations' share to 86 per cent of the world's recoverable reserves. Again, however, this exaggerates the importance of the arid lands since at least half of these reserves are in the non-arid portions of the nations.

In summary the basic geological structures of the arid lands, in particular the lack of the Hercynian Remnants (Table 4.1), means that coal production and reserves are not a major resource despite the somewhat misleading evidence of Table 8.4.

Petroleum
For the core arid and predominantly arid nations, it has been the discovery

Plate 8.1 Oil drilling rig, east of Kokand, USSR, photographed 1976. The site is on the southern slopes of the Ferghana basin above the level of the irrigated areas. The electric power lines are a reminder of the considerable development of thermal electricity as well as oil in central Asia over the last two decades.

111

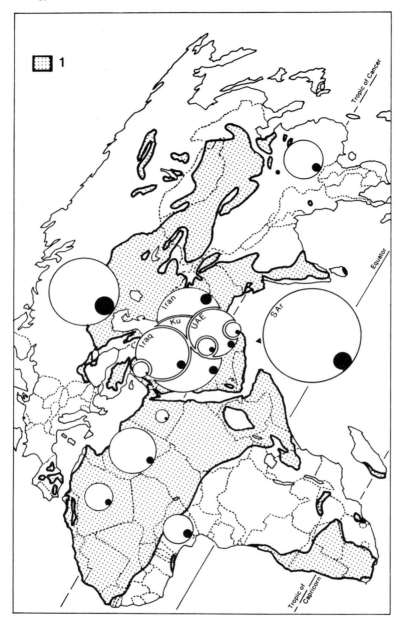

Fig. 8.1 Oil production and reserves in arid nations *c.* 1977. *Sources*: UN *Statistical Year Books*.

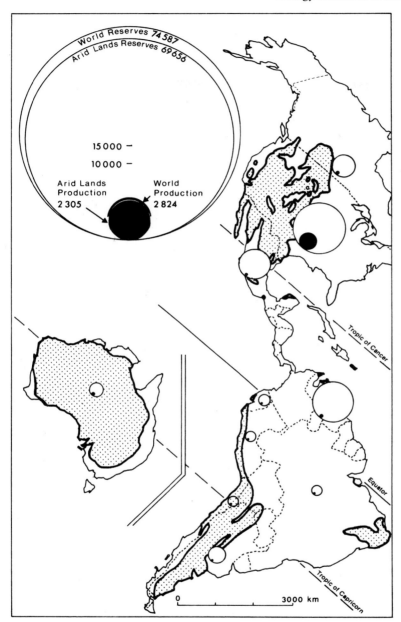

Key: 1 – Arid lands. Ku – Kuwait; SAr – Saudi Arabia; UAE – United
Arab Emirates

Production (average 1974–7) is indicated by solid circles; reserves (1977)
by open circles, each proportional to volume (expressed in millions of
tonnes).

113

Table 8.5 Global petroleum production 1860–1975

Date	Production (10^6 tonnes)	Share (%) produced by		
		USA	Near and Middle East	Others
1860	0.07	97	0	3
1880	4.1	89	0	11
1900	20.5	43	0	57
1910	44.9	64	0	36
1920	95.9	87	0	13
1930	197.2	66	2.5	31.5
1945	366.0	60	7	33
1955	769.0	46	18	36
1965	1512.0	32	27	41
1974	2870.0	22	59	19
1975	2650.2	15.5	41.2	43.3

Sources: *Information Zur Politischen Bildung*, Bonn, **162**, 1975, 21, and UN *Statistical Year Books*.

of the large petroleum reserves in the 1930s and 1940s which has transformed their economic and political status in the world (Pl. 8.1). The trends of global oil production lead from the initiation of major production in the USA to the dominance of the Near and Middle East nations by the mid-1970s (Table 8.5). In the late 1970s the core and predominantly arid nations provided 45 per cent of the global production and held 64 per cent of the reserves (Table 8.4). With the production of the semi-arid and peripherally arid nations the total arid lands' share ran to 82 per cent and to 93 per cent of the global reserves. Unlike the figures for coal, the bulk of these data does refer to production from and reserves in the arid lands and, when mapped does provide a reasonably accurate view of the role of the arid lands in this source of fossil fuel energy (Fig. 8.1). Indeed, a survey in 1979 suggested that of the four largest oilfields in the world, three were in the core arid nations (Saudi Arabia, Kuwait, and the Neutral Zone) and these with the fourth (Venezuela) represented 21 per cent of the global reserves (Dick and Marden 1979).

Natural gas
In terms of natural gas the arid nations' role is very similar to that for petroleum. The production from the core and predominantly arid areas, however, is much less than for petroleum, 6 per cent compared with 45 per cent and the share of the reserves is only about half that for petroleum, 36 per cent compared with 64 per cent. Yet total and nation production and reserves provided almost identical proportions of global figures as for petroleum, 81 per cent and 90–93 per cent respectively. The production from the USA (4932 million tetracalories), the USSR (2889 million tetracalories) and Canada (685 million tetracalories) is mainly responsible for the massive con-

tribution of the semi-arid and peripherally arid nations, and in this case the bulk of this production is coming from the arid portions of the various nations. In terms of reserves the USSR dominates with 22 billion tonnes, followed by Venezuela 11 billion tonnes, Iran 10.5 billion tonnes, USA 6 billion tonnes and Algeria 3 billion tonnes with Qatar, Nigeria, and Kuwait all over 1 billion tonnes. Again the bulk of the reserves seems to be in the arid lands.

Future energy management

It is one of the ironies of global geography and geology that those countries best endowed with solar flux energy, particularly as radiation, are also among the best endowed in terms of geologically stored solar energy in the form of fossil fuels. Thus, countries which had the greatest potential to capitalize on their radiant energy sources have been so well provided for by the industrialized world's discoveries of their fossil fuel reserves from the 1930s and 1940s onwards that the alternative flux sources have not needed to be developed.

For the immediately foreseeable future the arid nations, particularly those core and predominantly arid nations clustered around the Persian Gulf, will continue to dominate the global production of the most important fossil fuel and energy source – petroleum. Their wealth will continue to accumulate and their political allegiances continue to be the goal of super-power strategies.

At remote locations far from fossil fuel supplies, solar and wind energy will continue to be valuable low-cost alternatives but they lack the mobility of the petrol can.

PART III

Evolution and impact of human resource use in the arid lands

While the scientists, each in their specialist way, have provided assessments of the various resources of the arid lands, it has been obvious from the discussion of those assessments in Part II that not only have the assessments varied among the scientists, but that the realities of human use of the resources may have differed considerably from the views of the 'experts'. We need to examine the origins and the contemporary nature of resource use in the arid lands in order to explain, if we can, the differences in resource use around the world and the contrasts between the optimal use in the view of the experts and the actual use of resources by the arid lands population – the effective managers of the resources.

CHAPTER 9

Resource use strategies in the arid lands: an overview of past and present

Although details vary, most definitions of resources identify them as 'things' useful to mankind. Such things will vary from soils to minerals, from rainfall to climate, from empty space to the view of a landscape.

In assessing the evolution of resource use in the arid lands we need to keep in mind a series of questions about that use, which if applied consistently, should enable the significance of different resource uses to be assessed and compared. We can then consider the various hypothetical strategies for arid lands resource use as background to the overview of changes in resource use in the arid lands over the last 400 years, i.e. the period of the expansion of European culture and technology around the world. With this as introduction we may then consider the details of the various types of contemporary resource use in the arid lands and attempt an assessment of their success in providing a livelihood for the arid lands population.

Some questions on the nature of resources

Given the general definition, four kinds of questions can be posed for any system of resource use:

1. What is the character of the resource itself?
   - Is it ubiquitous or localized in space?
   - Is it continuous or restricted in time? Can it be used only once, several times, or an infinite number of times?
   - Can it be used completely or only partially?
2. What is the character of the use of the resource? This, in part, asks whether the use is matched to the character of the resource.
   - Is the use ubiquitous or localized in space?
   - Is the use continuous or restricted in time?
   - Does the use exhaust the resource or allow only partial use?

- Is the use of the resource voluntary or involuntary – the result of free choice or lack of any alternative?
- Is the use in isolation or linked to other resource use systems?
3. What is the population carrying capacity of the resource use?
   - How many people can be supported directly or indirectly per unit area by the use of the resource?
   - What standards of living can be provided for those people by the resource use?
   - To what extent do the people supported by the resource use control the use of the resource itself?
4. What evidence is available of differing appraisals of the resources?
   - Is there any evidence of cultural variations in resource appraisal?
   - What appears to be the strategy or philosophy behind use of the resources?

While answers to these questions will not always be as easy to provide as the questions are to formulate, some attempts at answers need to be made if the existing resource uses are to be evaluated.

Some strategies, philosophies, and models of resource use

Strategies

Historical attitudes to the global arid lands seem to fall into three groups. The first group recognizes the arid lands exist but ignores them as a resource base. Evidence of such views might come from place-names such as the Arabian Rhub al Khali – 'The Empty Quarter', or the existence of vacant unused or 'waste' lands – officially owned by the State but not allocated for use. In Australia, for example, such lands represented 27 per cent of the continent in the 1960s and still 18 per cent of the continent in the early 1980s (see Table 16.2).

The second group of attitudes recognizes the arid lands as a potential resource but allocates the use to others to develop and manage for the actual owners. Such a policy was implicit in the creation of political buffer states on the edges of the arid lands in southwest Asia and North Africa at the time of the Roman Empire and can be seen currently in the Great Power conflicts in the same area. Such a policy also is implicit in situations where the owners of the land allow others to search for and extract resources from the arid lands under some form of lease or licence system. Implicit here is the assumption that the actual owners either lack the means or the incentive to risk capital in the research for possible resources, but are willing to let others take the risk in return for a share of any profits. Such a policy existed in the 1900s–1960s with regard to oil exploration in southwest Asia and North

119

Africa. Although modified by subsequent nationalization of some of the oil companies, the principle is still implicit in most mineral exploration of the global arid lands in non-centrally planned economies.

The third group of attitudes recognizes the resources of the arid lands and reflects the determination of the owners to use the resources themselves. In some cases, for tribal groups driven out from the better lands by stronger tribes, there may have been no alternatives, and the use of the arid lands is in that sense involuntary. In most cases, however, the resource use is voluntary, a deliberate choice by the landowners to tap known potentials for themselves.

Philosophies

Use of the available resources might be characterized as either in relatively unmodified form or only after the resource has been substantially modified by processing or by supplementation. Such a division represents two ends of a spectrum of varying combinations of the two use systems, but the basic division needs to be borne in mind.

When resources are used in a relatively unmodified form, the assumption is that use can only occur when and where the resources become available, when and where the opportunity for use presents itself. This 'opportune use' is implicit in the nomadic pastoralists' use of the arid rangelands, their flocks and herds being moved to harvest the fodder as and where it occurs. Consumption of the resource is usually on site with the minimum of transport from source to demand. The demand (e.g. livestock) is moved to the sites at times when the supply (fodder) is available, and the fodder is consumed there, *not* harvested for stall-feeding later. Such opportune use relies on mobility in time and through space to search out resources, rather than a sedentary base to which resources are brought for consumption or storage.

At the other end of the spectrum of use the resources may be considerably modified before use. Such modification might be processing to ensure preservation over time (of say food or fodder) or supplementation of some resource deficiencies to take advantage of surpluses or other resources. Solar radiation might be excluded from human living quarters to reduce heat stress, but on the other hand may be concentrated in solar furnaces to provide industrial energy. Irrigation in the arid lands takes advantage of the high solar insulation, and through imported water offsets the local deficiencies to create a viable production system.

Models

While there are varying systems of resource use, there are also various models of actual use which imply varying philosophies towards the resources themselves. A review of the history of human use of global resources, not merely in the arid lands, reveals three models. Most definitions usually distinguish between 'fund' and 'flow' resources, between those which appear to be non-renewable and therefore of finite dimensions, and those which appear to be renewable (by biological processes) within useful time spans and therefore in theory infinitely available. Bearing this in mind, the three models can be identified as a *Depletion Model*, a *Conservation Model*, and an *Accretion Model*, each with a series of possible scenarios (Fig. 9.1).

The Depletion Model assumes rates of use faster than any renewal of the resource and leads to the rapid depletion or exhaustion of the resource, which itself leads to human concern or distress. Destruction of the resources might be virtually instantaneous as in the nuclear blast scenario (Fig. 9.1A1). A steady reduction of the resources to zero (Fig. 9.1A2) might be found in continuous mining of a particular ore, although in fact complete exhaustion even here might never take place because the efficiency of ore extraction rarely reaches 100 per cent. Alternatively, the increasing costs of extraction of the lower-grade ores over time might bring mining to a halt (Fig. 9.1A3) until the price of ore or cost of mining change the profitability of the operations. In another case a slow rate of initial use may accelerate because of the environmental impact of resource use on the host ecosystem (Fig. 9.1A4). An example could be the cumulative effects of dry-farming on the soils of the Great Plains of the USA and the local collapse of the ecosystem in the 'Dust Bowl' of the mid-1930s (see also Ch. 18).

The complete exhaustion of the resources appears to have been relatively rare in human history so far – the extinction of the American carrier pigeon being one exception. However, continued or increasing demands for non-renewable resources through greed, ignorance of the reserves, or a deliberate policy of exploitation in the shortest possible time, have created situations in the arid lands where the Depletion Model can be recognized. At such times the concern expressed regarding the trends of this type of resource use may lead to action which transforms the Depletion into the Conservation Model.

Where concern about resource depletion leads to action to slow down, halt, or reverse the trends of resource use, the Conservation Model can be identified (Fig. 9.1B). Again, various scenarios can be suggested. Each represents an attempt at management to conserve the resources – conservation – and usually implies attempts to reduce the rates of resource depletion by searching for alternatives, greater efficiencies in use, or even questioning the size of the demand for the resource.

When demand for the resource continues and depletion is inevitable, the rate of depletion may be reduced (Fig. 9.1B1) by more efficient use (e.g. more calories of energy from fossil fuels), or use of substitutes (e.g. alternatives to fossil fuels such as solar flux energy). Where depletion has to be stopped (Fig. 9.1B2), the resource may be preserved by restrictions on use (landscape reserves, national parks) or maintained by new replacement inputs (e.g. fertilizers to replace soil nutrients for agricultural yields). The resource levels may be brought back to original levels by sufficient inputs

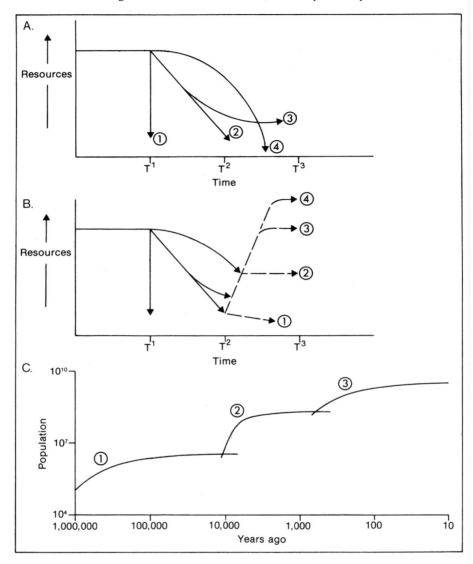

Fig. 9.1 Models of resource use. *Source*: Deevey 1971 and various
   Key:
   (A) *Depletion Model*: 1 – Instantaneous use/destruction of all resources,
   e.g. nuclear explosion; 2 – Gradual depletion through constant use, e.g.
   mining; 3 – Maintenance of reduced level after initial depletion, e.g. cessa-
   tion of mining; 4 – Accelerating rate of depletion, e.g. soil fertility under
   monoculture without refertilization.
   (B) *Conservation Model*: 1 – Reduced rate of use, e.g. by more efficient
   use; 2 – Preservation of remnant after use, e.g. reserves; 3 – Rehabilitation
   of resources to preserve levels by inputs from elsewhere, e.g. fertilizer ap-

(Fig. 9.1B3) or even built up to exceed the earlier levels (Fig. 9.1B4). The best examples here would be in agriculture where depleted farms have been re-established through fertilizer and capital inputs so that the biotic yield from the fields now exceeds the yield under the pre-agricultural ecosystem. Such instances usually imply, at least at the outset, a substantial transfer of resources from some other location as input into the devastated area. In the case of the Dust Bowl in the USA, as we shall see in Chapter 18, the reconstruction was achieved only by applying national financial resources to a depressed region of the nation. It is doubtful whether the region could have solved its problems by itself.

The previous models have implied general depletion of resources as a result of human use over time, but the third – the Accretion Model – (Fig. 9.1C) implies an improvement in availability of resources over time. Two types of phenomena might be interpreted in this way. The human population may be regarded as a rapidly increasing demand for global foodstuffs and raw materials. Yet to the entrepreneurs of those foodstuffs and raw materials that population represents an expanding market, and to the resource-use planners it represents an expanding supply of labour to manipulate the global environment for beneficial purposes. As 'mouths' and 'hands' the human population represents assets as well as liabilities. As 'brains' the human population represents the greatest asset of all – the ability to investigate and work out how to manipulate the global environment to best advantage. The graph of global population growth in the Accretion Model reflects in fact the quantum jumps of human population as the result of innovations in knowledge. While many fear the effects of human manipulation of the environment, there is no doubt that an increased ability to manipulate that environment has increased the theoretical range of available resources. It is also true, however, that while knowledge accumulates over time, it also may decrease or depreciate in value, whether as a result of technical obsolescence or changes in popular taste or culture. Thus the increases of the graph may be more apparent than real, and the accretions in fact be less than predicted. We need only remember the greater variety of foods available to the hunter–gatherers compared with our own. Nonetheless this and the two preceding models offer hypotheses against which to judge the realities of resource use in the arid lands.

---

plication in agriculture; 4 – Rehabilitation beyond prior levels, e.g. supply of trace element deficiencies in originally sterile soils
(C) *Accretion Model*: The accumulation of human knowledge expressed as various revolutions in technology increases the awareness of, and ability to use, new resources and thus enables the manipulation of the environment to support larger human populations. 1 – Cultural or Tool-making Revolution; 2 – Agricultural Revolution (the development of agriculture); 3 – The Scientific/Industrial Revolution

Decision theory

The realities of resource use also should indicate the extent to which the theoretical and actual range of resource-management choices coincide. Decision-theory research recognizes that there is a distinction between the theoretical range of choice in resource management – that range of adjustments and resource uses that have been practised in any similar environment plus a possible innovation, and the actual range of choices (White 1974, pp. 187–205). The actual range is believed to be smaller because of ignorance of the possibilities, problems of communication, preferred actions, and the constraints of a social, cultural, or religious nature. Further, the research identifies management *optimizers* who aim for the maximum possible benefits from resource manipulation, and management *satisficers* who are content with less than the optimal benefits provided they meet the modified goals.

The realities of the arid lands resource use will also demonstrate the problem of identification of the resource managers. Here they may be illiterate peasants struggling to provide a subsistence for their families with minimal capital or equipment and only the traditional methods of resource use as guidance; there they may be the directors of an international company with all the sophisticated communication systems at their fingertips, attempting to forecast trends in the global market and evaluating the chances for higher profits to placate their shareholders. A further group might be the central planners of the socialist states attempting to match actual production to goals set by political expediency or ideologies. Between these three groups of managers lie the differences of scale of thinking, scope for resource manipulation and incentives for action of the multitude of decision-makers. Yet each plays a role in the resource use of the global arid lands and each believes they are manipulating the arid lands as best they can.

Measuring the resources

While the question of identification of resources can be complex, even more difficult is the problem of resource measurement. Measures are only as good as the tools which make them and the measurers' perceptions of what ought to be measured. The geologists, faced by this problem, have devised a simple matrix to reconcile a spectrum of knowledge on the one hand with economic constraints on usefulness on the other. The matrix or McElvey Box provides a useful model for most exercises in resource appraisal and measurement (Fig. 9.2).

For the geologists a basic distinction is made between *reserves* of mineral ores – that is those ores which are known with reasonable certainty to exist and which are worth recovery at current prices, and those other mineral *resources*, the existence of which is uncertain (but which might be inferred

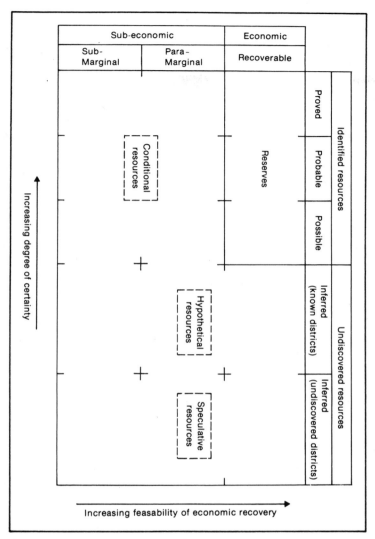

Fig. 9.2 The McKelvey Box. *Source*: Modified from McKelvey 1972 and Manners 1981

Note: 'Reserves' here are identified bodies of ore which are recoverable at current prices

'Resources' here are bodies of ore which are either known or expected (by extrapolation of existing knowledge) to exist or which are identified ore bodies too expensive to mine at current prices

'Possible' – at least one ore exposure known

'Inferred' – an exposure expected but no sample yet

'Paramarginal' resources would be recoverable at prices 1.5 times current levels from extrapolating existing knowledge) or which are known to exist but which are uneconomic to mine at the contemporary prices. The precise boundary between the possible and the inferred ores is the knowledge of the existence of at least one exposure of ore; one exposure upgrades the inferred to the possible. Between the recoverable and para-marginal the boundary is a price rise to 1.5 times the contemporary prices.

To the extent that this matrix acknowledges the interrelationship in the identification of resources (in the general sense) between gradations of knowledge and gradations of economic viability, it provides a useful model for most types of resource assessment. As we shall see, however, the economic criteria may be replaced by social or cultural criteria – preferences or taboos forming barriers to use of potential resources just as effective as those of cost.

Evolution of resource use in the arid lands, 1500–1980s

To understand both the patterns of contemporary use of the arid lands and some of the problems associated with them, we need a historical geographical viewpoint – a recognition that the patterns of the present often relate to the past and that present decisions on resource use have many historical precedents.

Although European contact with the arid lands of southwest and central Asia predates the end of the fifteenth century, it was the discovery of the Americas as part of the expanding power and interest of Europe (Parry 1963) which ushered in a new phase of arid land development – a phase which was to transform significantly both the use and condition of the global arid lands. Spanish colonization of central and southern America together with Portuguese colonization of Brazil brought European culture and superior technology for the first time into permanent contact with the arid lands. For almost 300 years thereafter, European experience of arid lands was limited to the experiences of these two nations in the Americas. The end of the eighteenth century, however, introduced a period when that experience was rapidly widened as settlers of the new republic of the United States, French settlers in North Africa, Dutch and British settlers in the Cape of Good Hope, and British settlers in Australia, began to occupy increasingly arid areas of the 'new lands' – whether as European colonies or newly acquired national territories. The changes in land use and resource management which resulted from this European invasion of at least the edges of the arid lands are suggested in Table 9.1 and Figures 9.3 and 9.4.

In AD 1500 only about 8 per cent of the arid lands appear to have been empty of human activity while hunters and gatherers ranged over about a quarter of the area in the Americas, South Africa, and Australia. Grazing of domesticated livestock was limited to the Asian and African arid lands with the full range of livestock, goats, asses, donkeys, sheep, horses, cattle,

Table 9.1 Arid land use *c*. AD 1500–1980

		Percentage of arid land area	
		c. AD 1500[a]	1980[b]
Unused (waste, uninhabited)		8	16
Hunting and gathering economies	N. America Patagonia Kalahari Australia ⎬ 23	Kalahari?	<1?
Livestock grazing economies:			
Nomadic pastoralism		50	41
Sedentary ranching		<1?	25
Rain-fed agriculture:			
Subsistence crop and livestock			
Shifting cultivation	⎬ 17		⎬ 12
Cash cropping			
Irrigated Agriculture		2	2
Woodland (use uncertain, 1980 only)		?	4
Total		100%	100%

[a] Interpolated from Bobek 1962 (and Fig. 9.3)
[b] Interpolated from *Goode's World Atlas*, 15th edn, Rand McNally, Chicago, 1978, and Fig. 9.4. Mining, tourism and military uses not included in above. Arid lands as per Fig. 2.1.

and camels in use. The bulk of the grazing appears to have been of the nomadic type and only a minute proportion was tied to fenced grazing areas. Rain-fed agriculture included the main cereal grains (wheat, barley, and millet) which were limited to Asia and Africa. The plough was not in use south of the Sahara in Africa or in the Americas or Australia. Large areas of rain-fed agriculture were in fact areas of shifting agriculture with periods of cultivation followed by long periods when no crops were grown or harvested. As a result, the area thus mapped as rain-fed agriculture therefore on Fig. 9.3 is probably much too large. Approximately 2 per cent of the arid lands were irrigated, but the systems varied considerably in size and technical sophistication. In all cases, however, they required a considerable manpower input to maintain the facilities.

In terms of the socio-economic systems of resource use, the range was from unspecialized hunters and gatherers through nomadic pastoralist societies, to sedentary rural civilizations based on rain-fed agriculture and urbanized 'rent–capitalist' societies based upon irrigated land or sedentary farming lands.

By the 1980s, the area of waste land appeared to have doubled, apparently as a result of the withdrawal of the hunter–gatherers and some of the nomadic pastoralists (Fig. 9.4). The total area grazed, however, had increased significantly, particularly that area where grazing was relatively

127

Fig. 9.3 Land use in the arid lands *c.* AD 1500. *Sources*: Modified after Bobek 1962

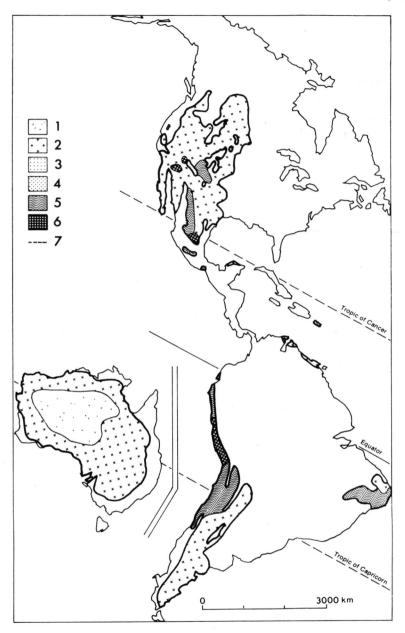

Key: 1 – Uninhabited waste lands; 2 – Hunting and gathering; 3 – Nomadic pastoralism; 4 – Sedentary ranching (none?); 5 – Rain-fed agriculture; 6 – Irrigated agriculture; 7 – Southern limit of plough usage

Note: Because of the nature of the sources the patterns are generalized and probably wrong in detail

129

Fig. 9.4 Land use in the arid lands *c.* 1980. *Source*: After E. P. Espenshade (ed.) (1978) *Goode's World Atlas*, 15th edn, Rand McNally, Chicago

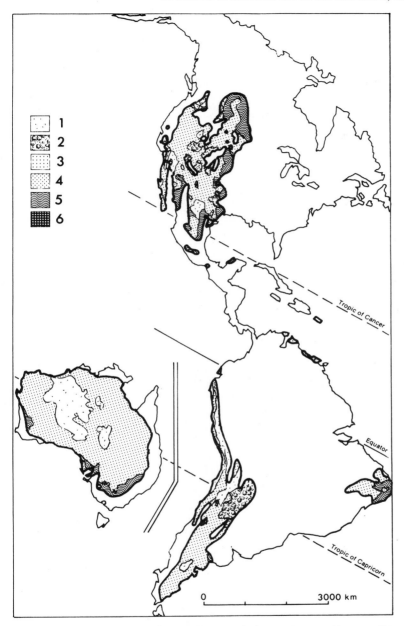

Key: 1 – Uninhabited waste land (unused except for possibly military or tourist purposes); 2 – Woodland (use uncertain); 3 – Nomadic pastoralism; 4 – Sedentary ranching; 5 – Rain-fed agriculture; 6 – Irrigated agriculture.

Table 9.2 Arid nations land use c. 1978

Land use (million ha)

	Arable	Permanent crop	Arable + permanent crop	Permanent pasture	Forest + Woodland	Other land	Total
	1	2	1 + 2	3	4	5	6
Group I (100% arid)	5.3	0.3	5.5	163.8	27.8	351.2	548.3
% of group area	—	—	1.0	29.9	5.1	64.0	100
Group II (75–99% arid)	145.2	6.4	151.6	853.1	273.3	1199.5	2 477.5
% of group area	—	—	6.1	34.4	11.0	48.5	100
Group III (50–74% arid)	77.9	14.8	92.7	441.3	109.0	96.5	739.5
% of group area	—	—	12.5	59.8	14.7	13.0	100
Sub-total Groups I–III	228.4	21.4	249.8	1 458.2	410.1	1647.2	3 765.3
% of group area	—	—	6.6	38.7	10.9	43.8	100
(% of world land use)	17.2	24.4	17.7	46.2	10.1	37.0	28.8
Group IV (25–49% arid)	490.1	10.3	500.4	661.8	723.2	849.5	2 734.9
% of group area	—	—	18.3	24.2	26.4	31.1	100
Group V (1–24% arid)	354.2	20.2	374.4	745.4	2025.0	1515.9	4 660.7
% of group area	—	—	8.0	16.0	43.5	32.5	100
Total arid nations	1072.6	52.0	1124.6	2865.4	3158.3	4012.6	11 160.9
(% of world land use)	80.9	59.1	79.5	90.9	77.9	90.1	85.4
World total	1326.3	87.9	1414.2	3150.9	4056.7	4451.8	13 073.6

Source: FAO Year Books.

fixed in space, i.e. sedentary, and now occupied two-thirds of the arid lands. This implies a considerably increased pressure upon the grazing resource, particularly where domestic livestock (cattle, sheep, etc.) had not been present in 1500. Rain-fed agriculture had decreased in area especially in North Africa. The main regional increase, associated with the transfer of the plough technology, was in the semi-arid areas of the Americas, southern Africa, and Australia. Irrigation areas had remained in approximately the same proportion but again there were significant local oscillations of area.

Land use in the arid nations in the 1970s

The information offered by the FAO *Year Books* for the arid nations (Table 9.2) provides four basic categories of land use (cropland: arable and permanent, permanent pasture, forest and woodland, and 'other land' including waste and urban uses). Combining the data for the three most arid groups of nations (Groups I, II, and III), virtually 44 per cent of their areas are unused, while almost 40 per cent are in pasture, 11 per cent in woodlands and only 7 per cent is cropped (and most of that not every year). The importance of this pasture land globally is shown when the total area of these nations (some 29 per cent of the world's area) is compared with their share of the global pasture lands, namely 46 per cent. Similarly, they contain 37 per cent of the unused lands of the world.

Comparing data for the first three groups, the substantially arid nations (Group III) have the lowest unused area (13%), the largest proportion in pasture (60%), in woodland (15%) and cropped (13%). Conversely, as might be expected, the core arid nations (Group I) have the smallest proportion of their area as pastures, woodlands, or cropped, and have almost two-thirds (64%) of their area unused. The Group II nations' data fall between these two extremes.

For the semi-arid and peripherally arid nations (Groups IV and V) the land use reflects the larger areas of humid climates with woodlands occupying from 26 per cent to 44 per cent, larger proportions cropped (8–18%), and smaller proportions grazed (16–24%) and unused (31–33%). The environmental constraints on land use in the arid nations were still obvious in the 1970s.

The transformations hinted at by these admittedly crude global measures reflect transfers of people, plants, and animals through space and over time. Some of those transfers of plants in particular were, as we shall see, accidental, but their effects were none the less important and occasionally disastrous. On the global scale the changes were intensifications of the pressures on the arid land resources by both an invading alien population and an expanding indigenous population. The complexities of these changes are examined separately in the following chapters.

CHAPTER 10

The nomads

Nomadic use of resources implies the harvesting of naturally occurring surpluses as and where they occur, either directly by human labour or indirectly through the use of domesticated livestock. Such a system requires knowledge of the appearance of those surpluses and the ability to move to them at the right time and in sequence to maintain a constant supply for consumption. Although such a system of resource use could be self-sufficient, more usually the lack of particular foodstuffs, tools, or other desired items had to be made up by barter from other resource users.

In terms of the extent of the resources used and the relevant specialist techniques for their use, two types of nomadic resource use can be identified: the hunter–gathering peoples and the pastoral nomads. In the 1980s the areas of the arid lands used by these two groups are but a fraction of that area of even 100 years ago (Table 9.1). The total population solely engaged in nomadism in the arid lands is certainly less than 13 millions and more likely to be less than 5 millions, or between 0.8 per cent and 2.1 per cent of the arid lands' population. Yet the systems of resource use have relevance today, not only because they are still practised, but also because they require techniques and strategies of resource use which seem to be particularly suited to the characteristics of some of the arid resources. Furthermore, their systems of resource use appear to have had a significant impact on the contemporary arid landscapes.

The hunter–gatherers

[At least a million years] before the onset of agriculture man had evolved with his present form, and all the basic patterns of human behaviour had appeared: language, complex social life, arts, complex technology. . . . If we are to understand the origin of man, we must understand man the hunter and woman the gatherer.

(Washburn 1976: xv)

Origins

The earliest evidence of man in the arid lands are relics of hunting–gathering peoples, such as the buffalo (*Bubalus* – now extinct) hunters of the central Saharan Tassili Mountains of 6000 BC (Lhote 1960). Since they were dependent entirely upon the naturally occurring surpluses and their management systems were limited to seeking out and harvesting those surpluses, they were not particularly well adapted to the arid or extreme arid ecosystems and preferred the richer resources of the more humid lands. The archaeological evidence, mainly from southwest Asia, suggests that the actual hunting–gathering system was replaced by agriculture and/or domestic livestock grazing in association with agriculture. The ancient hypothesis widely held until the mid-nineteenth century that pastoral nomadism was the half-way step from hunting–gathering to agriculture seems to have been disproved by evidence of livestock domestication as part of the development of agriculture (see Kramer 1967). That domestication, according to Childe (1952), was itself a reaction to the desiccation of the plains of southwest Asia at the end of the Pleistocene period (Baumhoff 1978). A switch to agriculture was the answer to increasing aridity.

The success of agriculture, the argument goes on, led to the creation of pastoral nomadism as a specialist subsystem of agricultural peoples:

> As the numbers of people and stock increased, herdsmen and herd moved farther and farther away from the villages and became more permanently detached from the settled lands. This about describes the roots of pastoral nomadism. (Sauer 1952: 97)

It has been described as 'a specialized offshoot of agriculture that developed along the dry margins of rainfall cultivation' (Johnson 1969: 2).

There is still debate about the order and rationale for this sequence and not all archaeological sites tell the same story. The Tassili hunter–gatherers for example were replaced by cattle herders about 3500 BC with no evidence of agriculture (Lhote 1960). Yet the fact that the hunter–gatherers do not seem inherently adapted to the arid environments makes surprising their occupation of almost a third of the area in 1500 and some of the most remote areas today.

The explanation for these locations seems to lie in their unsuccessful conflict with invading groups having different resource-use systems. The hunter–gatherers appear to have been driven out of their better environments to areas which were relatively unattractive to their persecutors. This appears to have been true for the Bushmen or San people now living in the Kalahari Desert and for those Australian Aborigines who are returning to traditional hunting–gathering systems on the reserves and granted tribal lands of the central Australian deserts.

The nomads

Relics

The San peoples

In the case of the San peoples of southern Africa, persecution from Hottentot goat herders from the south, Bantu cattle herders from the east, Herero cattle herders from the north, and Ovambo agriculturalists also from the north (in present-day Angola) (Tobias 1964), together with a 'systematic Dutch extermination campaign' after their colonization of the Cape of Good

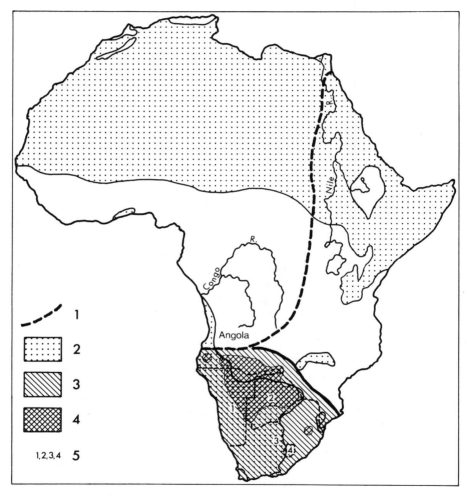

Fig. 10.1 The retreat of the hunter–gatherers in South Africa. *Source*: Tobias 1964
Key: 1 – Western limit of skeletal finds about 10,000 years BP; 2 – Arid lands; 3 – Occupied by hunter–gatherers AD 1652 at time of Dutch settlement of the Cape; 4 – Occupied by hunter–gatherers, 1960; 5 – 1 = Namibia; 2 = Botswana; 3 = Republic of South Africa; 4 = Lesotho

Hope in 1652 (Lee and De Vore 1976: 5) forced them to retreat to the sands of the Kalahari (Fig. 10.1). Of the 50 000 San people remaining, only 5 per cent or 2500 are still practising hunting and gathering. The remainder are mostly working for their erstwhile black or white persecutors as a wage-labour force on the invading cattle ranches, or as paramilitary forces patrolling the desert borders of Namibia or Botswana, or in some cases as agriculturalists squatting on their traditional tribal lands.

The Australian Aborigines
In the case of the Australian Aborigines, the retreat of the hunter-gatherers in the face of persecution by British settlers and governments also has been documented (Rowley 1972). The bulk of the remaining tribal reserves lie within the arid zone (Fig. 10.2). The majority of the current Aboriginal population of 161 000 (1976) no longer rely upon hunting-gathering but have become either a wage-labour force on cattle and sheep stations, or an urban workforce, or – for the less successful – urban fringe dwellers relying on unemployment benefits and welfare handouts from governments. The 1970s saw a revival of interest in traditional life styles, however, resulting from a revived racial feeling supported by several groups in the white Australian community. The outcome was a demand for land rights to the tribal reserve areas and much of the as yet vacant Crown lands of the arid interior. Effective control of these reserves is passing into Aboriginal hands through the creation of Aboriginal-dominated governing bodies and in 1981 the South Australian Government granted the first Aboriginal inalienable freehold title for a large arid land reserve (some 10% of the State's area) to the Pitjinjatjara Aboriginal people.

Part of the land rights agitation was for homelands from which to derive cash income via mineral royalties or grazing activities, but part was also a desire to return to traditional hunting-gathering strategies, albeit with use of motor vehicles and modern weaponry. This 'outstation movement' has resulted in the movement of Aborigines from established settlement centres, whether pastoral station or towns, back to the 'homelands' where camps have been established around watering points and attempts made to re-create traditional life styles.

The Alice Springs area case study
In 1978 about 8000 Aboriginals lived within the Alice Springs Pastoral District, of which 1000 lived in or around the two main towns (Fig. 10.2). The remainder were spread on government reserves (60% of population), Aboriginal pastoral leaseholds (10%), or non-Aboriginal pastoral stations (20%). As part of its policy as a newly created State (1978) of the Australian Commonwealth, the Northern Territory Government set in motion procedures for reviewing Aboriginal claims to vacant Crown land. By January 1978, 20 per cent of the Alice Springs District was under claim and up to 35 per cent could be liable to such claims. Even on the government reserves, although the financial support still comes from official sources central control

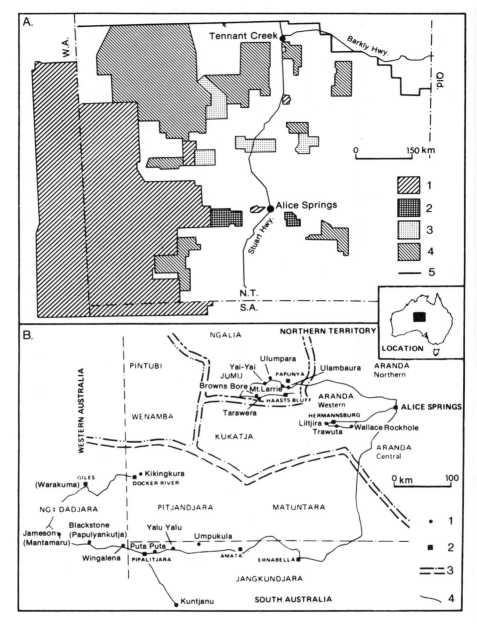

Fig. 10.2 Aboriginal settlement in central Australia *c.* 1979. *Source*: Griffin and Lendon 1979

Key: (A) Central Australia showing land tenures relevant to Aborigines. 1 – Aboriginal reserves; 2 – Special mission leases; 3 – Pastoral leases leased by Aborigines, 4 – Areas under claim by Aboriginal groups, 5 – northern boundary of Alice Springs Pastoral District.

(B) Aboriginal settlements and tribal language boundaries. 1 – Aboriginal outstations; 2 – Main settlements; 3 – Tribal language boundaries; 4 – Survey route.

of affairs on the reserve is in Aboriginal hands – usually by a council or congress of tribal and family representatives relying upon traditional lines of descent and authority. On these reserves the outstation movement has led to groups moving away from the central government settlement to set up new daughter settlements along more traditional lines (Griffin and Lendon 1979).

A survey of some of these settlements in mid-1978 found varying degrees of success – whether judged in terms of economic self-sufficiency or social and psychological satisfaction with the experiment. Although some hunting and gathering were practised, the main activities were subsistence gardening, some cattle grazing, and some mining. The return to traditional life styles applied, if at all, more to the social relationships than the resource use and the practical problems facing such settlements were considerable (Table 10.1).

The lost resource-use system

In the face of the pressures from competing resource uses the hunting–gathering life collapsed or was pushed back into the remotest and least hospitable ecosystems. From the point of view of resource-use theory, however, the systems of survival which the hunter–gatherers had evolved had much to recommend them. Recent research has provided evidence that their systems were remarkably efficient.

The efficiency of the systems reflected the wide range of potentially acceptable food supplies, intimate knowledge of the occurrence of those supplies over space and time within territories, a highly mobile society, and kinship linkages which assisted the groups to survive periods of environmental stress (Fig. 10.3A). Most researchers have stressed the omnivorous preferences of the Kalahari peoples and similarly the Australian Aborigine (see Ch. 7.3). Up to 50 per cent of the available species of plants and animals were regularly eaten by Kalahari groups in the 1960s. Detailed knowledge of seasonal water supplies and seasonal ecological conditions above and beyond those now known was reported in all studies of hunting–gathering societies. Material possessions were minimal – weapons and tools, food and water containers, loincloths, and perhaps a skin cloak, with perhaps a few decorative beads. The campsites were clusters of temporary wind-shelters of branches, leaves, and bark around the fireplaces (Tobias 1964; Lawrence 1969).

Population densities varied with the available resources. In the Great Basin of the arid western USA prior to European settlement, a Shoshone family of five needed 300–400 km² of territory, or up to 80 km² per person, but the Paiutes needed only 5 km² per person. Kalahari bushmen needed 7 km² per person while Aboriginals in central Australia probably approximated the Shoshone figures (Daryll Forde 1950; Lawrence 1969). Yet there is apparently no evidence of hunter–gatherers ever starving to death. This

Table 10.1 Problems facing Aboriginal outstations in central Australia in 1978

Basic problem	Immediate expression	Future expression
A. *Technical problems*		
1. Establishment of new settlement site	Lack of information on water quantity and quality Delays in establishing wells and bores Lack of knowledge of horticulture and inadequate advice	?
2. Maintenance of the settlement once established	Maintenance of bores Inadequate stores/supplies Inadequate transport and communications Inadequate education services Few employment opportunities Inadequate housing	Lack of official knowledge of Aboriginal language and social organization Management of cattle grazing Increasing energy demands as population grows Co-ordination of official support services Few and temporary official advisers
3. Environmental deterioration	Removal of vegetation for timber and fuel	Management of grazing lands
4. Competing uses		Pressure for access from tourist and mining industries
B. *Non-technical problems*		
1. Aboriginal society	Conflict in authority between traditional and modern spokesmen (e.g. elders versus official trainees)	?
2. Culture clash	Conflict in value systems between Aboriginal and non-Aboriginal resource-use systems.	?

Source: Griffin and Lendon 1979.

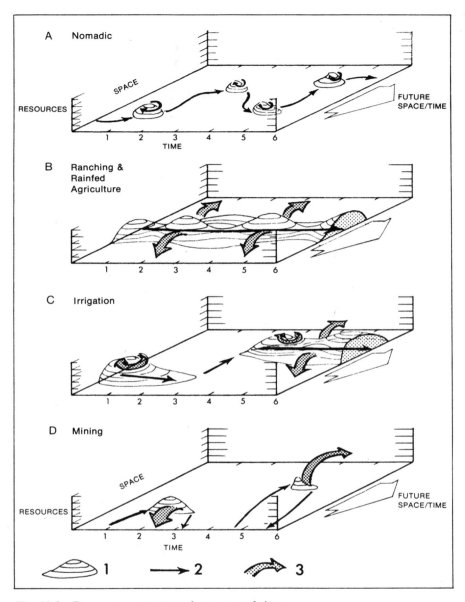

Fig. 10.3 Resource-use systems in space and time

The four diagrams attempt to show the sequence and intensity of resource uses under four different systems: A – Nomadic; B – Ranching and rain-fed agriculture; C – Irrigation; D – Mining

Key:

1 – Cumulative isopleths of intensity of resource production at locations in time and space

2 – Paths of resource users moving from one area which has been exhausted of its resources to another offering further resources

3 – Export of produce out of the immediate resource production system

141

raises both the question of whether they practised population control and the question of the energy balance of their activity.

On the question of population control there is documented evidence for manipulation of birth-rates and infant survival by a variety of measures to reduce the frequency of intercourse, some evidence of contraceptive measures and delayed weaning of infants, and evidence of loss of life by abortion, infanticide, senilicide, invalidicide, and inter-group fighting, particularly at times of environmental stress. When one adds to this the complexities of clan and kinship ties which require the sharing of all food and water supplies – a social organization which was most complex where the rainfall was lightest, most seasonal, and least reliable (Coon 1971) – we can appreciate the resilience of this resource-use system.

The system is not only resilient to stress. It provided time for leisure and creative cultured activity. R. B. Lee's studies of Kalahari groups have suggested that provision of calories from hunting and gathering was more than adequate for the energy expended. Two-thirds of the tribal group were able to provide all the food requirements for the group in just over a third of the available time. He concluded that although they had only a very simple technology 'yet they lived quite comfortably in a severe and unsparing place' (Lee and De Vore 1976).

Environmental impact

The sparse population densities and limited interference with the arid ecosystems would seem to indicate a very limited impact of the resource-use system on the environment. Certainly in terms of the harvesting of plants and animals it is doubtful whether any major long-term change in conditions resulted. Two possible exceptions, however, need to be mentioned.

All hunter–gatherers in the arid lands seem to have made use of fire either as a hunting weapon – to drive game into a line of hunters, or as a means of luring game to the new vegetation which colonized the burnt areas. European explorers in Australia often commented upon the burnt areas they had crossed or reported Aborigines firing the vegetation as a deliberate management strategy. Similar uses are reported in the Kalahari and by the pre-European groups in the arid Americas. The effect of such fires in the arid environment must have been considerable, since once lit they were rarely deliberately extinguished. Depending upon whether the accumulation of ground litter and vegetation allowed a 'hot' or a 'cool' burn, trees may or may not have been killed, and among the grasses and herbs the annual species probably favoured at the expense of the perennials. Such practices may have affected the arid ecosystems over several millennia but their impacts are impossible to estimate precisely. Nevertheless, such stresses must have had a significant role in the extension of grasslands at the expense of woodlands on the margins of the arid lands, especially when a period of higher rainfall allowed the build-up of annual vegetation as potential fuel.

A more contentious impact is that of the hunting process upon the world's megafauna which disappeared at the end of the Pleistocene era about the time when evidence of agriculture began to appear. On the one hand, the demise of the large game animals is seen as the challenge which forced the hunter–gatherers to become agriculturalists (Martin 1967). On the other hand, it is argued that the activities of the hunters *led* to the extinction of the animals and left gaps in the ecological niches, in particular in the global semi-arid to sub-humid grasslands.

The pastoral nomads

Reviewing the state of nomadism on the Arabian Peninsula in 1968 Peppelenbosch (1968: 339) suggested that the phenomenon was best described as

> non-sedentary animal husbandry determined by the search for pastures . . . [where] an entire human group accompanies the flocks and herds . . . [in a] seasonal . . . but . . . fixed round of movements . . . [around] permanent wells. . . . The nomadic society is not self-sufficient.

As with the hunting–gathering form of nomadism this type of resource use was more extensive 100 years ago, but now occupies about 41 per cent of the arid lands (Table 9.1). Even this figure is an overestimate as it does not take into account the recent trends towards the settlement of pastoral nomads in permanent bases.

The contemporary pastoral nomads

Estimates of the numbers of pastoral nomads can rarely be based upon censuses since few exist and the essential character of the resource use often implies the crossing of national frontiers as part of the grazing movements. In the 1950s–1960s various estimates, all of questionable accuracy in detail, suggested a total of 20 million pastoral nomads (Table 10.2). This figure is too high since it includes groups which were nomadic for only part of the time and uses only the highest estimates available. North Africa, including the Sudan, had 5 millions, East Africa 1.7 millions, southwest Asia 7 millions, and central Asia 6.2 millions. In terms of being nations of nomads, the Sudan and Somalia (although then still colonial or mandated territories) and Mongolia came closest, with up to 40 per cent of the Sudan population identified as nomads in the mid-1950s (Davies 1966) and 74 per cent of the Somali population likewise identified in 1959 (Silberman 1959). In the early years of this century Mongolia had up to 80 per cent of its population as pastoral nomads, but the proportion has been declining since the 1950s with 'modernization' of the State's economy.

Table 10.2 Pastoral nomad populations *c.* 1960s

Area	Population estimates (*000*)
1. East Africa[a]	
Masai (Kenya)	60.6
Masai (Tanganyika)	46.0
Mukogodo	3.3
Somali	640.0
	(1 480)[b]
Turkana	80.0
2. North Africa and southwest Asia	
Sahara[c]	~1 000.0
Arabia[c]	~1 000.0
Turkey, Iran, Pakistan[c]	4 000–5 000.0
India[d]	1 000.0 ?
Sudan[e]	4 080.0 ?
3. Asia	
Soviet central Asia[f]	5 510.0
Sinkiang[f]	670.0
Total	19 929.9 (includes highest estimates in all cases)

Notes and sources:
[a] Allan 1965: 307. These are his 'Pastoralists' category.
[b] Silberman 1959.
[c] Dresch 1966.
[d] Bhimaya 1960. (200 000 families × 5).
[e] Davies 1966. His estimate of up to 40% of population of 10.25 million.
[f] Freeberne 1966.

Johnson (1969) classified the patterns of traditional movements of people and livestock in pastoral nomadism into two basic types: vertical movements, where the seasonal shift was related to the availability of pastures at different

Fig. 10.4 Pastoral nomad movements. *Sources*: A–D Johnson 1969; E modified after Cordes and Scholz 1980: 41
Key: (A) – Horizontal (elliptical), e.g. Ruwala of Arabia; (B) Horizontal (pulsatory), e.g. Tuareg of North Africa; (C) Vertical (constricted oscillatory), e.g. Kurds of Iran/Iraq; (D) Vertical (Limited amplitude), e.g. Teda-Tibesti of North Africa; (E) Routes of bedouin in Dhafrath Region of Abu Dhabi Emirate
Key for diagrams A to D: 1 – Oases; 2 – Wadis; 3 – Normal routes; 4 – Good year routes, abundant forage; 5 – Bad year routes, scarce forage; 6 – Mountains
Key for diagram E: 7 – Pearl fishing boat movements (early summer offshore; late summer coastal); 8 – Movements of Bani Yas tribal livestock; 9 – Movements of Manasir tribal livestock; 10 – Date gardens: open = private; solid = common property; 11 – Winter grazing areas; 12 – Summer grazing areas; 13 – Bad year movements

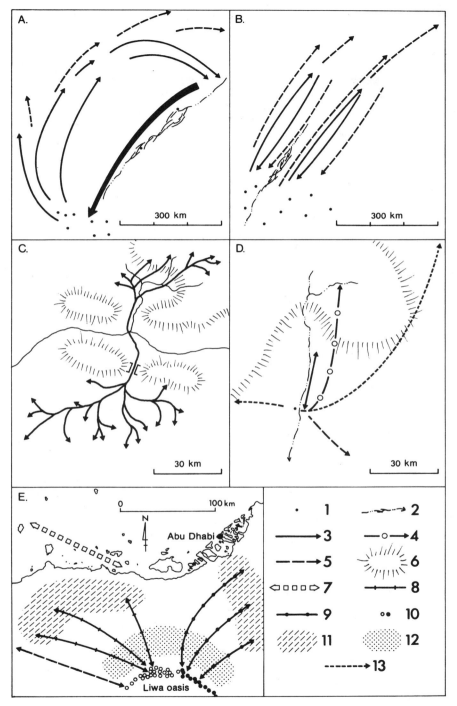

altitudes with transfers to and from plains and mountains, and horizontal movements, where the shift was to pastures scattered over plains or rolling country. Various modifications of the types were indicated and the dimensions of the movements were contrasted. The vertical movements tended to be from one-tenth to a quarter of the distances travelled in horizontal movements (Fig. 10.4). Johnson noted that the camel was not used on the vertical movements, the dominant livestock being sheep, goats, and cattle with horses as personal transport, whereas the horizontal movements included camels, horses, and goats with both camels and horses also serving as personal transport.

This basic classification hides a wide spectrum of variations not only of movements but of the dependence of the groups upon their livestock. Many groups have preferred bases – usually summer camps around permanent water – which may be agricultural villages inhabited by either tribal members or related social or cultural groups. The complexities of the movements are obvious from Fig. 10.4 and the Abu Dhabi case involves groups which earlier combined pearl diving and agriculture with nomadic livestock movements.

Sedentarization policies

The 1960s–1970s saw an increasing tendency for nomadic pastoralists to settle at particular locations on a permanent or almost permanent basis. This process of sedentarization was the result in part of the development policies of the newly independent arid nations of Africa in particular, but also the result of independent historical trends related to the opportunities for, and attractions of, sedentary life. These latter need to be explained first.

Historical evidence from North Africa and Arabia shows that nomadic tribal groups may not always have been either nomads or pastoralists. Competition between tribes for territory produced a kaleidoscope of environments available to the tribes, which seem to have responded by adapting their resource-use systems to the changing possibilities. T. E. Lawrence (1963: 35) speculated that the conflicts between agriculturalists in the more humid highlands of Yemen spawned pastoral nomads:

> The congestion of the Yemen . . . [forced] the weaker aggregations of its border down and down the slopes of the hills . . . towards the deserts of Nejd. These weaker clans had continually to exchange good springs and fertile palms for poorer springs and scantier palms, till at last they reached an area where a proper agricultural life became impossible. They then began to eke out their precarious husbandry by breeding sheep and camels. . . . Finally, under a last impulse from the straining population behind them, the border people (now almost wholly pastoral) were flung out of the furthest crazy oasis into the untrodden wilderness as nomads.

He noted the process still operating at the time of the First World War and from the evidence of place-names in the Widian Valley suggested this area had been the 'factory of nomads, the springing of the gulf-stream of desert wanderers' who had ranged as far as the frontiers of Syria and Mesopotamia. A similar process was documented for the El Gwasim tribe of Morocco, which in 1907 consisted of farmers and sheep breeders, but by 1934 had become only camel breeders and practised no agriculture (Monteil 1959).

It would seem that for the arid and extreme arid areas pastoral nomadism may have been a refugee resource use – the result of a limited choice forced upon refugee groups pushed out of the richer semi-arid ecosystems. If so, then the current apparent reversal of that expulsion, that is the nomads leaving their grazing areas for alternative livelihoods as urban or agricultural labourers, may be understood to be a return to a life style of a not-so-remote past. With oil company wages offering in 20–40 days what took a year to accumulate even working in the oases, the economic incentives to remain a pastoral nomad are being eroded rapidly.

Part of the disincentive to remain pastoralists has come from the erosion of their political and military power. The success of Lawrence's camel-borne nomad guerrillas against the Turkish Army in the First World War would be impossible to emulate today, since air power, off-road vehicles, helicopter gunships, and wireless communication have destroyed the nomads' monopoly and control of desert space. Their social status has been eroded also, as a deliberate policy of newly independent, urban-oriented, agriculturally and industrially based nations. Thus the nomad Tuareg's traditional control of their slave labourers in the Saharan oases was removed by the independent Government of Algeria, and the aim of most such governments has been the conversion of their nomads to sedentary ranchers or agriculturalists (Pl. 10.1).

From Egypt to Mongolia the pressure on the nomads accumulated in the 1950s and has been increasing steadily through the subsequent years. Lattimore (1962: 186) reported in 1962 that:

> the Mongol planners are aiming to convert the livestock economy from a nomadic structure to something like a ranch structure, with most of the people sedentary most of the time.

Communal ranches of 2000 people and 56 000 horses were to be created around a semi-permanent village with hospital, meeting hall, and service centre. Similar ranching schemes in southwestern Angola were running into difficulties in the early 1970s, both from the environmental degradation of livestock concentration and from the necessary disruption of traditional socio-economic patterns (de Carvalho 1974). In the Sudan, sedentarization schemes appear to have reduced the nomadic component of the national population from the mid-1950s to 1973. The 1955–6 Census identified 1.4–1.8 million nomads (some 14–18% of the population), although Davies (1966) suggested 3.1–4.8 millions (some 30–40%). In the 1973 Census only 1.6 million nomads were recognized – only 11 per cent of the total popu-

Plate 10.1 Backyard views in state farm, 'The Way of Lenin', near Ashkhabad, USSR, photographed 1976.

The state farm was established in 1968 to provide cotton crops with water from the Kara Kum Canal which reached Ashkhabad (50 km to the west) in 1962. The prior population of nomadic Turkmenis was supplemented by Russian technicians and within 8 years the treeless steppe had been converted to cotton fields surrounding the main settlement of 5400 people. While the nomads had become agricultural workers living in brick houses with access to television, bicycles, and health services, in their backyards they maintained a few vestiges of traditional life, namely camels (top) and the occasional yurt (bottom).

lation. Again, ranches had been set up as pilot projects but again, they had been virtual failures. This had been the result of planning from above not below; of the annexation to the pilot ranches of the best communal grazing grounds leading to the disruption of traditional systems which had balanced these areas against the poorer lands as well as denying use of these better lands to the majority of pastoralists; and of the attempts to reduce the live-stock numbers held by the pastoralists as part of range-management policies (Khogali 1979). Better success had come from settlements of pastoralists on the edge of the agricultural areas where the traditional integration of use – grazing of stubbles in return for the livestock manure and a share of the produce – could be reinforced. Hopefully, the new measures would not result in the same disasters as were reported for sedentarization schemes in Turkmenia in the 1930s, when forced settlement and livestock reductions of 80–90 per cent led to local famines and loss of life (Monteil 1959).

Evaluating the policies

And the land was not able to bear them, that they might dwell together: for their substance [flocks, herds, and followers] was great, so that they could not dwell together. And there was strife between the herdmen of Abraham's cattle and the herdmen of Lot's cattle.

(Genesis 13: 6–7)

While there seems to be no dispute about the existence of sedentarization policies, there is considerable debate about their positive and negative effects – a debate which harks back to the conflict between the herdsmen of Abraham and Lot. The positive effects of sedentarization are seen as the reduction of grazing pressures on the arid ranges when the nomads' livestock are withdrawn. All pastoralists appear to have the belief, based upon long practical experience, that wealth and social prestige are increased in pro-portion to livestock owned, and traditionally the automatic increase in live-stock by unlimited breeding was slowed down or reversed only by diseases, warfare, or natural disaster such as drought. Indeed the suggestion has been made that for the Bedouin:

the taking and retaking of camels through raiding seems then to . . . [have served] as a means of circulating or distributing a scarce resource over a wide area.

(Sweet 1965: 142)

After the Sahel drought of 1968–72 the reaction of a Fulani herder in the spring of 1973 was: 'I had 100 head of cattle and lost 50 in the drought. Next time I will have 200' (Glantz 1976a). The sentiments are common to all tra-ditional pastoralists. Livestock are the famine insurance, the more the better the insurance. Yet the effect of unlimited increases in livestock numbers are obvious in the environmental effects of overgrazing. The pressure on resources must have been the occasion for hardship and conflict even before it confronted Abraham and Lot.

Many commentators see such pressures and conflicts as inevitable and suggest that the system of resource use is inherently unstable:

149

> Nomadic pastoralism is inherently self-destructive, since systems of management are based on the short-term objective of keeping as many animals as possible alive, without regard to the long-term conservation of land resources.
>
> (Allan 1965: 321)

In so far as the removal of the pastoralists' livestock will remove this pressure on the environment, the demise of nomadic pastoralism in this context is seen as a benefit. Further benefits, it is argued, may accrue from the pool of labour so freed for other employment.

Some commentators see the adoption by the nomads of the camel as an environmental disaster:

> the camel, by enabling man to lead a nomadic existence, a less socialized life, has made it possible for him to exploit the vegetation over ever vaster and more varied stretches of land; so that the camel is responsible for the creation of the desert.
>
> (Charles Sauvage, botanist, quoted in Monteil 1959: 573)

Further, several of the African pastoralists regard an overgrazed range with its sparse vegetation as less of a disease hazard for their livestock than lush pastures and are therefore not concerned about apparent reduction of vegetation from overgrazing (Allan 1965).

On the negative side, the replacement of nomadic grazing by sedentary grazing usually runs the risk of increasing livestock pressure on a constricted area of range unless a reduction of livestock (which will always be resisted by the pastoralists) is effective. The record of such schemes seems to show more failures than successes.

Withdrawal of nomadic grazing may leave vast areas of arid land even more unproductive than before, since at least the nomadic use obtained some sustenance from such areas through the livestock grazing. On the remote ranges, the grass will still grow whether or not there are livestock to eat it, and no other harvesting mechanism yet devised offers as efficient a method as livestock for such remote and low productive areas.

The end of nomadic pastoralism would also mean the end of detailed knowledge of the grazing resources of large areas of the arid lands, and since for a large proportion of the lands no other resources have yet been identified, this may prove to be a serious long-term loss. French geographers noted that Saharan nomads often had long records, either written or orally recorded, of places where pasture had been found. The earliest records were dated 1671–1744 for the Brabish tribe, but several tribes had records for over 50 years and some for over 100 years from the eighteenth century onwards (Monteil 1959). As historical evidence of possible climatic fluctuations reflected in the vegetation as well as guides to the chances of fodder at particular locations, such records are invaluable. The oral records in particular will disappear with the last of the nomads and human knowledge in general will be that much the less.

150

CHAPTER 11

Ranching

Comparing the patterns of nomadic land use with sedentary pastoralism or ranching over the period from AD 1500 onwards, it is obvious that ranching has taken over large areas of the arid lands which were previously either hunting–gathering or nomadic pastoral economies. Over large areas of the arid lands in the Americas, southern Africa, central Asia, and Australia the domestic livestock have occupied the semi-arid and arid climatic zones, introducing into these areas often for the first time intensive and regular grazing pressures on the environments.

Ranching is a related but intensified form of nomadic pastoralism. It implies similarly the rearing of ruminant herbivores, the products of which are for commercial sale and usually exported out of the production area. The livestock are also basically dependent upon live feed which they search out and harvest themselves, but in contrast to the nomads this is on a fixed permanent range owned directly by an individual, partnership, or company. Further, the operations of the production system are directed from a permanent base on the range itself.

Global location: theory and practice

In terms of its global location, ranching is peripheral to the sedentary agricultural resource use systems (see Fig. 9.4). Except where nomadic pastoralism is still practised further out into the arid lands, ranching is often the last form of resource use before the uninhabited waste lands of the arid core. On its arid borders are the empty lands, on its humid borders are the farmers. This peripheral location seems to be the result of a combination of circumstances, partly reflecting the resource potentials of the locations, but also reflecting economic pressures and the history of development of resources.

Based upon an analysis of the economics of transport, Von Thünen in 1826 suggested that provided the resource potentials were equal over space,

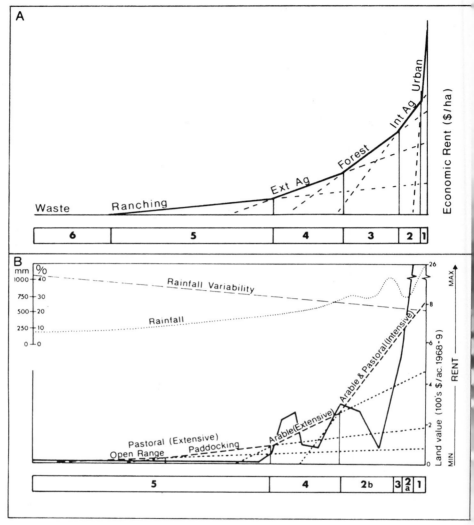

Fig. 11.1 Economic rent and land use: models and realities. A graph of economic rents as suggested by Von Thünen (A) is compared with actual land values and land use in Australia (B) and movement of meat products to the European market (C + D). *Sources*: A, B Heathcote 1977; C, D Crossley 1976, after Figs 1 and 2 respectively

Key: (A) Von Thünen's economic rent model 1826. Hypothetical land use was 1 – Urban industrial; 2 – Intensive agriculture; 3 – Forest; 4 – Extensive agriculture, 5 – Ranching, 6 – Waste land.

(B) Land use and land values (free market sale prices) in New South Wales in the 1960s. Hypothetical bid-rent cones for land use from Sydney to the South Australian border using the Von Thünen model are indicated by dashed lines (from intensive land use near the Sydney market to pastoralism on unfenced 'open range' in the remote interior). Actual land use in the 1960s was: 1 – Urban (Sydney metropolitan areas); 2a – Intensive agriculture (horticulture); 2b – Intensive pastoral (improved pastures for wool, beef and mutton production); 3 – Woodland (mainly native); 4 – Extensive agriculture (grain crops); 5 – Extensive pastoral (sheep and cat-

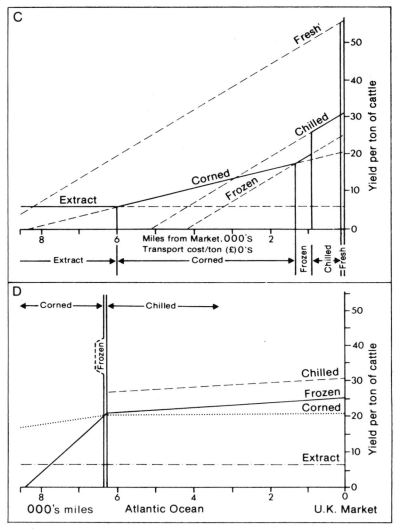

tle ranches on fenced range in paddocks)
Average rainfall (in mm) and variability (in %) along the section are also indicated
(C) Hypothetical zonation of beef product industries in the 1920s.
Yields of beef products per ton of cattle are indicated along a section from the London market, showing distances of 000s of miles, and transport costs in tens of £s per ton based upon arbitrary transport rates and speeds. Hypothetical yield cones are dashed lines, actual yield cones solid lines
(D) Actual zonation of beef production from the United Kingdom–River Plate countries of South America, 1924. Yields of beef products per ton of cattle are indicated against actual transport charges and journey times. The effect of cheap ocean transport was to extend the distance chilled beef could be shipped up to the 6300 miles limit which included the River Plate estuary. Inland from the estuary, however, transport costs rapidly removed the possibility of either chilled or frozen beef production and only corned beef was profitable

153

the existence of a market for produce at any point, say a town or city, would result in a series of concentric zones of different production systems around the market, differentiated according to the value and perishability of the product and the cost of its transport to that market. In the sequence of zones the grazing of domestic livestock on natural feed would be the most remote and last zone of productive use before transport costs prevented any use of the land further away from the market (Fig. 11.1). That grazing could be the remotest use was a function of the fact that the livestock could transport themselves to market at minimal cost. That there were limits to this movement resulted from the fact that driven animals lose weight and condition on the road (since they are burning up more energy each day than the feed is supplying) and the costs of droving would eventually be higher than the market return on the emaciated stock.

When further variables are added to the model, such as cheaper transport or increased production costs from climatic constraints, variations in the basic patterns can be forecast. Cheaper transport would enable more remote production than before, whereas climatic constraints might reduce the potentially productive areas. Since agriculture generally needs more precipitation than pastoral land use we would expect the agricultural zone to be at the

Table 11.1 Food transport from arid lands

A. Imports to Britain

Commodities	*Average distances moved by British imports 1830–1913 (miles)*		
	1831–5	*1871–5*	*1909–13*
Fruit and vegetables	0[a]	535	1 880
Butter, cheese, eggs	262	1 340	3 120
Meat and tallow	2 000	3 740	6 250
Wheat and flour	2 430	4 200	5 950
Wool and hides	2 330	10 000	10 900

[a] No significant imports.   *Source*: Peet 1969.

B. Meat processing and perishability constraints

Type of meat product	*Concentration (% of original animal retained)*	*Storage period possible (days)*
Fresh beef	61.0	~14
Chilled beef	61.0	~40
Frozen beef	61.0	>40 ?
Corned beef	25.3	∝ ?
Beef extract	1.5	∝ ?

Source: Crossley 1976. See also Fig. 11.1.

humid, and the pastoral zone at the arid, end of any precipitation spectrum. A decrease of average precipitation below 75–100 mm would probably preclude *any* pastoral use, nomadic or sedentary (Andreae 1966).

If we add a further variable, the spatial expansion of an alien population pushing out from a base in the humid area down the precipitation gradient into the arid areas, but producing for a remote market in the humid lands, we approximate the realities of European expansion into the semi-arid and arid lands of the Americas, southern Africa, central Asia, and Australia, particularly from the early nineteenth century onwards. The realities of global food production for the European market were noted by Peet (1969) and the details for beef production by Crossley (1976) (Table 11.1). Within Australia the spread inland of European settlement down the rainfall gradient from Sydney resulted in land values and land use which by the 1970s mirrored Von Thünen's model (Fig. 11.1). Model and reality were remarkably close on both global and regional scales.

Evolution of ranching in arid lands

To explain the contemporary location of ranching in the arid lands we need to combine the principles of Von Thünen's model with the historical geography of land settlement in the areas. The evolution of ranching in the global arid lands from 1800 onwards is the story of the invasion of the semi-arid and arid grasslands and shrublands by European settlers or their descendants, bringing with them domestic livestock and a system of livestock management which had been developed in Europe and only slightly modified before it was thrust into the arid lands.

The setting in 1800

In 1800 the bulk of the arid lands of the Americas were under Spanish jurisdiction, but effective control and Spanish settlements were absent from most of the arid areas of the present-day USA (with the exception of the string of missions along the Californian coast and the Rio Grande in present-day New Mexico), the Atacama Desert, and southern Patagonia. Horses, cattle, sheep, and goats had been introduced, but Spanish imperial trading policies and bans on merino sheep exports prevented the development of anything more than a restricted trade in animal skins with the home country (Parry 1963). Ranching had been established on free land grants from the Crown in Mexico and the South American arid lands and provided a source of local meat and raw materials. The original livestock and technology – the use of unfenced ranges, annual communal musters for marking and marketing of livestock, the basic equipment, and in particular the use of the horse as transport – had been imported from Spain, where ranching operations had been a 'pioneer' form of resource use on war-devastated lands conquered from the Moors in the period immediately prior to Columbus' discoveries.

In Australia in 1800, the British penal colony on the east coast was struggling to survive, its food supplies in doubt and its horses, cattle, and sheep (including the newly arrived Spanish merinos) finding the grazing of the humid woodlands thin and seasonally sparse. Of the interior semi-arid grasslands there was as yet no knowledge available to the intruders, and the hunting–gathering Aborigines were still masters of those lands, unaware of the animals and men soon to dispossess them.

In southern Africa the British had just obtained by treaty the control of the Dutch colony at the Cape of Good Hope. The Dutch settlers with a century and a half of experience of local conditions already had begun to push their livestock into the semi-arid and arid interior grasslands of the Karoo where from 1730 onwards they had met up with pastoral invaders, the Bantu, moving south and west with their zebu cattle. The Dutch, or Boers', livestock was by 1800 a mixture of European cattle and sheep with Hottentot long-tailed sheep and goats and the Bantu zebu cattle, but on the arid frontiers it was the sheep which were the pioneer stock. Although by 1800 Spanish merino ewes smuggled out of Spain via Britain had arrived here and some had been sent on to Australia to provide the foundation for the latter's fine-wool industry, the main commercial products of the European pastoral system in southern Africa were mutton and tallow for the victualling port of Cape Town (Guelke 1976; Talbot 1961).

In central Asia in 1800 the pastoral nomads were still in complete control, moving across the steppes and trading with the oases. As yet there were no hints as to the imperial designs for their future to be hatched in the capitals of Europe in the latter half of the century.

The Golden Age of ranching, 1830s–1920s

From the 1830s to the 1920s the European pastoralists pushed their livestock down the climatic gradient of rainfall into the arid lands. The semi-arid and arid grass and shrublands offered seasonal surpluses of feed broken by periodic droughts, and the natural competitors were soon either eliminated (e.g. the bison from the Great Plains) or their numbers so reduced as to be a minimal threat to the domestic animals (e.g. kangaroos in Australia and the deer and antelope of the South African veld). Diseases were relatively few and while carnivores such as the coyote in North America, jackal in South Africa, and dingo in Australia took their toll, only in Australia did control measures need a major capital investment and in this case – the dog-proof fences erected to protect the sheep-grazing areas in the latter half of the nineteenth century – the costs were borne by the central governments.

The indigenous human occupants of the potential rangelands were either killed or conquered and incorporated into the production system as a labour force, or removed to more remote reserves. The battles between Boer settlers and Bantu tribes, the bitter conflicts on the Great Plains between Red Indians and Anglo-Americans and on the plains of Patagonia between Indians and Argentinians, together with the less spectacular but equally decisive skirmishes between Aborigines and white Australians opened up the ranges, and the surviving conquered peoples often became shepherds and stockmen on the new ranches.

Expansion

Growing populations and improving standards of living in Western Europe during the latter half of the nineteenth century, resulting from a combination of longer life spans through improvements in public health and medical services, improved communications through railway linkages, and the capital surpluses generated by the Industrial Revolution, meant that demands for food (both bread grains and meat) were increasing over the period. In addition there was an expanding market for livestock products – even for a brief period in the late nineteenth century for South African ostrich feathers! The waves of European emigrants, at least 17 millions leaving Europe between 1850 and 1900, added to the local food demands in the eastern USA, South America, southern Africa, and Australia and provided a local stimulus for what rapidly became an international trading system.

The Boer 'Great Trek' of 1836 had pushed the European livestock around the fringes of the High Veld and the heart of the Karoo arid ranges. From the 1840s onwards the Australian pastoralists pushed in towards the central deserts and from the 1850s the American ranchers began to transfer their operations from natural grassland openings in the eastern humid woodlands to the tall-grass prairies of the Mid-West and short grasses of the semi-arid Great Plains. By the turn of the century, certainly by the time of the First World War, the world's ranchers had occupied most of the ranges which were to be occupied still in the 1980s. On the humid edge of this range they were to lose land to the advancing agriculturalists in the later decades of the nineteenth century, but on the arid edge they were to remain relatively unchallenged.

Resource exploitation

The apparent success of this pastoral invasion, as judged in terms of profits and the new population (both animal and human) which it supported, resulted from the combination of grazing resources made economically attractive by political actions and technological innovations. The political claims to the new ranges, whether by refugees from central authorities (as the Boers) or as part of government policies backed up by military force, opened up the *possibility* of new resource uses. The remoteness of these ranges, their aridity as reflected in seasonal shortages of feed and water, and their low productivity per unit area posed problems which only human ingenuity, technical innovations, and the expanding markets were able to solve.

The remoteness of the production areas from the markets was met initially by driving the livestock to market – a method in use on the Spanish plateau in Europe and on British 'drove roads' at least 300 years before the discovery of the new rangelands. Subsequently, from the 1870s onwards, the railroad networks were expanded to the edges of the range country in the Americas, South Africa, and Australia using the profits from earlier railroad construction in Europe or by government bond issues. Ocean shipping links, particularly using the slower but more reliable steamer services from the 1860s onwards, offered cheap freights to European markets. At the same time new techniques both reduced the costs of processing and prolonged the life of the product, enabling it to be longer *en route* to market.

Gideon (1948) has pointed out that the assembly-line principle of division of labour was first introduced into the abattoirs in Chicago in the 1860s to cope with the flood of cattle pouring in from the Great Plains and destined for the beef markets of the eastern USA and Europe. The first corned-beef patent taken out in 1875 further expanded the range of transport of the product, as did also, even more significantly in terms of its long-term effects on sales, the introduction of refrigeration processes during 1867–82 (Fig. 11.1C and D).

In terms of their life span, the products could now be shipped around the world to market (Table 11.1). A world-wide marketing system could not exist, however, without a communication system – to tell producers of market prices, for agents to co-ordinate their purchases and sales, and for carriers to organize their transports. With the world-wide network of telegraph lines and cables from the 1870s onwards, such a system became available.

Coping with aridity

Although never completely solved, the aridity problems were mitigated by sales of surplus stocks at times of stress; evacuation in a semi-nomadic fashion to any unoccupied productive range that was available; leasing of pastures from pastoralists in other areas; integration of irrigated hay meadows with open-range operations as in the southwestern USA; or if no other solution were possible, by letting the livestock fend for themselves and hoping that the next rains would find some survivors. Economically this latter was often the cheapest solution.

Water supplies were the main problem on the ranges from the 1830s to the 1880s. Over this period surface storages, created by damming creeks, excavating tanks to catch run-off from playas or in creek-beds, or sinking shallow wells by hand to be worked by horse-whims (a rotary windlass worked by a horse walking around it) had to suffice. In the 1880s, as a result of adoption of the deep-drilling technology which had been devised for oil searches, deeper underground supplies (some of them artesian) were discovered. The artesian waters were free-flowing, but simple windmills were adapted as pumps for the non-flowing bores. By the turn of the century, on most of the ranges, the deaths of livestock from the periodic droughts were no longer from thirst but starvation; all the feed within stock walking distance of the watering point having been eaten.

Remoteness also meant labour shortages, and loss of shepherds and stock-men to the world gold-rushes from the 1850s through to the 1890s (Morrell 1940) aggravated an already difficult situation. In part through this loss and in part as the result of pressure from the farmers for fencing laws to protect their crops from the marauding livestock, the ranges began to be fenced from the 1880s onwards. The need for constant herding was thus removed and the labour demands reduced in proportion. Mechanization of some of the ranching operations, such as shearing and pressing the wool into bales for transport, also became available about this time.

Remoteness may also have played a part in encouraging some companies to integrate their ranching operations with the processing side of the industry. Firms such as Armour and Star in Chicago, integrating Argentinian ranches and Chicago packing houses, and Vesteys and Bovril in Britain, managing Australian ranches for the British market, began to appear late in the nineteenth century and the global integration of livestock production and consumption from the arid rangelands to the European markets became a reality (Perren 1971). Thus, a heavy fall of rain on an outback Australian ranch with its promise of a good wool or cattle crop, when telegraphed to the Stock Exchange in London, could raise the price of pastoral company shares (Heathcote 1965).

Profits from the pastoral operations could be impressive. From 1879 to 1888 at least thirty-three companies were formed in Britain to invest funds in ranches around the world (Atherton 1961). By 1907 some £2620 million had been invested, 45 per cent in North America, 13 per cent in Australasia 18 per cent in Africa, 17 per cent in Asia, and the rest in Europe (*Quarterly Review* **208**, 1907). How much went to the arid lands can only be surmised but such funds, together with the surpluses from local gold mining, or in South Africa also diamond mining, no doubt provided some of the necessary capital.

The low productivity per unit area was met by systems of land tenure which provided sufficient land at cheap prices to enable the production units to be profitable. In the Americas, South Africa, and Australia, there seems to have been a common sequence: first, illegal occupation of the land at no cost followed by official regulation of the fact of occupation by a system of licences or leases at nominal or very low fees or rents.

In southern Africa after 50 years of Dutch settlement, part of it illegally on unpurchased lands, a system of grazing licences at fixed rents had been introduced in 1703. By 1812, however, it was reported that most pastoralists were in fact using at least 75 per cent more land than their legal title allowed (Duly 1968). Original practice had produced units of 2500 ha – being a circle of half an hour's horse ride radius from the central headquarters, but after the Great Trek of 1836 pioneer pastoralists were allowed two such units, each often up to 3200 ha, and again in practice many ranches were twice as big again (Christopher 1976).

The Spanish system of large free land grants in the Americas was the basis for the large pastoral and agricultural haciendas which had appeared by 1800. The revolutionary wars of the nineteenth century changed the ownership but not usually the resource-use system of such holdings. In the USA the pastoralists used the agriculturally oriented land laws after the Homestead Act of 1862 to acquire only the strategic water points on their ranges. They continued to use the intervening dry rangelands of the unallocated federal lands as free range until 1934, when the Taylor Grazing Act forced them to accept nominal grazing fees and control on stock numbers and grazing areas for the first time (Gates 1968).

In Australia illegal use (squatting) was followed by licences from 1835, and pastoral leases from 1847 onwards were seen as but the first of an escalating scale of intensity of resource development, culminating in land purchase for agriculture. For the arid lands such hopes were never realized, the only purchases being of the strategic water points as in the USA. In general there was no limit on the size of holdings until the 1880s and then only for certain types of lease and only in certain of the colonies, later states of the Commonwealth (Heathcote 1977).

The end of an era

The expansion of the ranching operations seems to have peaked about the time of the First World War. Since about the 1920s the profitability of the resource-use system has declined relatively, although still providing significant local employment, private profits, and national export incomes for the countries involved.

The decline of relative profitability seems to have resulted from a combination of circumstances. Over a period from the late 1880s in the USA to 1914–18 in southern Africa and Australia, a series of disasters hit the pastoral industry, particularly on its arid fringes. In the winter of 1887 a blizzard on the high plains of the western USA decimated the cattle and sheep wintering on the ranges. Almost overnight several pastoral companies lost their 'capital' – the livestock – and were bankrupted. In southern Africa the Boer War (1899–1902) disrupted pastoral production as much of the fighting was across the semi-arid countryside. From 1895 to 1901 in eastern Australia a series of droughts halved the livestock numbers. Many companies were bankrupted.

Further droughts during 1914 in Australia and southern Africa led to public concern for the future of the industry and the pastoral resources. What appears to have happened is that the original pastoral resources had been so modified by the increasing grazing pressures of the previous 20 to 50 years that occurrence of droughts had greater impacts than before. The combination of more livestock at risk because of increased permanent water points, their removal of the grassy components of the vegetation ('eating the haystack' as one Australian pastoralist ruefully put it), and the erroneous belief that no long-term damage was being done to the resources led to disaster. In the USA, the problem was in part a result of erroneous optimism, based on limited prior experience, that livestock could tolerate the bitter winters of the semi-arid high plains.

At the same time as these events were occurring pressure grew on the pastoralists to give up the better watered ranges to the farmers, who were at their heels in the search for new lands for development. Part of the reason for the Boer War had been the threat to the Boer pastoralists from the new British immigrants, originally as miners but also as farmers. In the USA federal land policies favoured the farmers on small 65 ha (originally 160-acre) homesteads. Several western states passed fence laws requiring the pastoralists to fence *in* their livestock rather than the farmers having to bear the cost of fencing them *out* (Gates 1968). In Australia similar legislation encouraged the spread of small farms as 'Closer Settlement' in preference to the extensive ranching operations, while officially sponsored 'Soldier Settlements' set up demobilized soldiers (after 1918 and 1945) as farmers on the better-watered lands compulsorily purchased from the pastoralists (Williams 1974). In retreat before the farmers' advance, the pastoralists were forced on to remoter more arid ranges and, when the next droughts came, disaster struck.

Ranching in the 1980s

Although changed in details the ranching systems in the 1980s retain many of their characteristics from the late nineteenth century. They are still the remotest areas under continuous resource use, and are still providing income as profits or wages from lands which would otherwise offer no returns. Their livestock still depend mainly upon the natural vegetation on the arid ranges and their produce still is exported to markets thousands of kilometres away, either as live animals for fattening or slaughter, or as by-products such as wool and skins which can tolerate the long distance and duration of transit. The ranch headquarters are still miniature service centres in their own right, with stores and repair facilities for mechanical equipment, fuel stores for the transport media (now oil fuel for motor vehicles rather than the hay and oats for horses), and dormitories or houses for the labour force. Additions will be the wireless aerials for local or regional communications and television aerials for social and cultural entertainment, together with an airstrip and aircraft for use as part of ranch surveillance and maintenance operations and as private long-distance transport.

The production units

Ranching in the 1980s takes many forms, from the experimental fixed bases of the sedentarized nomads in Mongolia, Sudan, and North Africa through

161

Table 11.2 Arid ranch production units *c.* 1960s

Criteria	Australia[a]	Western USA[b]	Namibia[c]	South Africa[d]
Ranch size (ha):				
cattle ranch	441 680	4000	—	3305–3584
sheep ranch	23 430	4800	800–2400	—
Cropped area (%)	0	~2	~1	~1
Livestock:				
(sheep	0.2–0.3	3–4[e]	0.1–0.2	0.1–0.2
equivalent units	(cattle and	(cattle and	(sheep)	(cattle)
per ha)	sheep)	sheep)		
Ownership	Family and	Family and	Family?	Family
	company	company		
Land tenure				
(% ranch)				
leasehold	99%	60%[e]	—	—
freehold	1%	40%	100%	100%
Capital investment:				
$ ha^{-1}	0.9–7.3	37.0–42.0[e]	?	~ 3.0
total ($)	170 000–409 000	150 000–200 000	?	~ $30 000
Labour	Cattle 10–15?	Cattle 5		
(man/years)	Sheep 3–4	Sheep 2+	10–18	~ 8
Product	Breeding cattle;	Breeding and fat	Karakul	Breeding and
	wool and cull	cattle and wool	pelts and	fat cattle
	sheep		wool	

Notes and sources:
[a] Heathcote 1975; 100.
[b] Clawson *et al.* 1960.
[c] Logan 1960 and 1973.
[d] Oberholzer 1957, data for the Molopo Region on the southern border of Botswana.
[e] The USA data do not distinguish between ranches in arid and humid areas, hence the livestock densities and investment figures are probably too high. It is also difficult to establish if the areas of public domain grazing annually leased are included in the area used.

the Karakul ranches of Namibia to the sheep and cattle ranches of Australia and the Americas. Recent comparative data at the scale of individual production units are difficult to obtain, so the comparative table setting out the characteristics of these units only shows the conditions in the early 1960s (Table 11.2). At that time the size of the ranches ranged from 800 to over 440 000 ha, from the smaller family-operated properties of the Molopo District of South Africa, employing perhaps half a dozen native workers, through the family sheep properties ('home maintenance' or 'living areas') of eastern arid Australia employing perhaps 2–3 workers, to the massive company-owned cattle stations of Western Australia. Here a manager and 9–14 workers, including possibly some Aboriginal stockmen, provided the labour force, with each worker in theory responsible for some 1200 cattle on 30 000 ha.

Livestock densities were generally low, reflecting the virtual complete dependence upon naturally occurring feed on the range. The American ranches, however, usually provided supplementary feed from some cropland and the higher densities there may have reflected this. Capital investment generally paralleled the size of properties, but was considerably higher in the USA. While there may be some exaggeration in these figures, they probably do represent a greater investment per unit area in 'improvements', such as fencing, watering points, access roads, and some machinery for the small cropped areas.

The produce from the ranches is both live animals for breeding or slaughter and their by-products as skins or wool. The economic success in the past and to some extent still in the 1980s has come from economies of scale of production, freedom from a heavy burden of overhead costs, and a relatively efficient labour force. The production system has benefited from its very simple (in many ways primitive) character, for example the annual rates of cattle sales ('turn-off') from central Australia in the 1970s were from 9 per cent to 19 per cent of the herd – little different from the rates for the Masai herds of Kenya according to a visiting expert (Box 1973). The investment per unit area has been kept low so that the inevitable losses from seasonal fluctuations of feed supplies have been somewhat minimized. In the light of this fact, proposals to invest large sums in development and further 'improvements' on the arid rangelands will need to be assessed carefully. The line between under- and over-capitalization is a very thin one on the arid ranges.

Ranching in the national economies

Attempting to separate the arid ranching component from national ranching operations is difficult, since in most cases the systems are integrated and data are not separated out in the literature. Crude estimates of the arid ranching components for the USA and Australia are given in Table 11.3. In the USA in the 1960s over half the beef cattle and three-quarters of the sheep were located in the western states and a significant proportion of these would be dependent during the winter upon semi-arid or arid basins as at least two-thirds of their feed came from the ranges. This was supplemented by feed from irrigated crops such as hay or alfalfa in the summer, when grazing was also in the humid highlands, possibly on reserves in the National Forests. In Australia the share of the national livestock and their produce was less, more like a third of national totals.

The environmental impact of ranching

The destructive effects of the concentration of domestic livestock on the arid range of South Africa were noted briefly in Chapter 7, but the evidence of

163

Table 11.3 Role of arid ranching production in the national economy of Australia and the USA *c.* 1960s

Criteria	Percentage of national figures		
	Australia		USA
Livestock:			
cattle	35	(all cattle)	46
		(beef cattle)	57
sheep	30		74
Production:			
beef	~30		41
veal	10		36
lamb/mutton	10		54
wool	33		75
Population (total arid area)	3		26

Sources: Australia – Slatyer and Perry 1969; USA – Hodgson 1963 (data for 17 western states).

ranching impacts can be drawn from many sources.

In Australia the effects of 'overgrazing' were noted by Royal Commissions into the pastoral industries of Queensland in 1897 and New South Wales in 1901 and there was further concern for 'desertification' in the 1930s (Heathcote 1965; Ratcliffe 1963). At the same time in the western USA, illegal grazing of federal arid and humid rangelands was blamed for massive and accelerating destruction of the rangeland resources (USA 1936).

A common sequence of deterioration seems to be evident. First came the removal of the edible species and thereby the encouragement of inedible species. Then, if pressures continued, the modification of soil and water conditions to the extent that all vegetation was removed and soil erosion became a major problem. Associated with this deterioration of the vegetation were rapid reductions in livestock carrying capacities, usually by massive deaths in droughts. In Australia the average reduction in grazing capacities from the peak numbers of the 1890s 'Golden Age' to the estimates of safe carrying capacities after 1920 was 50 per cent. In the USA overgrazing was thought responsible for carrying capacities in the 1930s being only 48 per cent of those in the 1880s.

Reduction of grazing capacities was not merely the result of destruction of arid vegetation. It also could result from the spread of relatively inedible species such as the mesquite tree and creosote-bush of the western USA, the *Acacia* thorn scrubs in South Africa, and the pine scrubs in eastern Australia. Part of the spread of these woody species seems to have been related to the pastoralists' concern to prevent wild (bush) fires destroying the grass cover. As suggested earlier, since many of the areas occupied by pastoralists in the nineteenth century were natural grasslands previously inhabited by hunting–gathering peoples who seem to have used fire fairly regularly in their resource-use system, the *reduction* of fires under the pastoralists may have allowed tree seedlings, which were previously regularly killed by fires, to survive to maturity. Further, the reduction of the grass cover through grazing would reduce the heat intensity of any further fire and more seedlings would therefore survive.

Not all the impacts were detrimental. There is evidence at least from Australia that wild animals benefited from the extension of permanent watering points in the arid ranges. Birds and larger mammals including kangaroos appear to have used the new waterholes to spread regularly into

Table 11.4 The condition of the arid rangelands (Australia and USA)

A. The USA

	Percentage of range at dates		
	1930–5	*1955–9*	*1975*
1. Forage condition:[a]			
Improving	1	24	19
Unchanged	6	57	65
Declining	93	19	16
	100%	100%	100%
Total area (million ha)	51.7	61.5	65.8
2. ·Soil erosion condition:[a]			
Stable			9
Slight	?	?	46
Moderate			35
Critical			9
Severe			1
			100%

B. Australia[b]

	1975
Soil conservation needs:	
Area not requiring treatment	44.8
Area requiring management practices only	22.3
Area requiring conservation works	32.8
Area requiring change of use	0.1
	100%
Land area in use (million ha)	335.6

Sources: [a] Hadley, 1977. Data for grazing districts in 11 western states.
[b] Australia 1978: 146. Data for arid grazing area = 43.7% of total national land area.

165

country previously used only seasonally. Again the impression is that the grazing pressures on the environment were increased as a result, but the general impact is very difficult to estimate. There is no doubt, however, that this development did not ease the pressure on the environment.

The environmental impact of arid lands ranching has continued since the 1920s and there is evidence of the varying magnitude and seriousness of that impact. Russian researchers have reported that the sand ridges and salt-pan (*takyr*) plains of the Kara Kum Desert (where rainfall averages 148 mm annually) need from 7 to 17 years between grazing to allow natural regeneration of the shrubs and low tree forms (ICASALS 1978: 5). Trends in the western USA since the 1930s seem to have shown some improvement in range condition by the 1950s and 1970s, although in terms of soil erosion 45 per cent of the range still had moderate to severe effects evident (Table 11.4A). In Australia a national survey in 1975 gave evidence that about one-third of the arid grazing areas required major soil conservation works and a further fifth needed improved management to reduce soil erosion (Table 11.4B). The survey reported:

> The arid grazing areas generally have the lowest priority in terms of value of production and fixed assets relative to the cost of soil conservation works. As the measures required for treatment in the degraded arid areas typically include some destocking measures, the implementation of conservation programs is likely to result in even lower returns – at least in the short term. While priority of treatment in economic terms is low, action is urgent since rehabilitation in arid areas is difficult and requires considerable time to reach full effectiveness.   (Australia 1978: 137)

However, at the time of the survey and in the years following to 1980 actual livestock numbers were increasing in central Australia as a result of the combination of wetter years (in 1975 the rainfall was over three times the average!) and poor prices for livestock, discouraging pastoralists from marketing their annual surpluses. The result by 1980 was that the area around Alice Springs was carrying some 550 000 cattle on ranges which had carried only 130 000 at the end of the last major 7-year drought of the 1960s. Apart from demonstrating the capacity of the range in good years, the figures boded ill for the survival of the permanent vegetation in the next drought. The environmental impact of pastoral resource use on the arid lands has been obviously considerable in the past and is continuing apace.

CHAPTER 12

Rain-fed agriculture and marginal land use

Global patterns

From AD 1500 to the present day, rain-fed agriculture seems to have declined relatively in area, from 17 per cent to 12 per cent of the arid lands (see Table 9.1). Such crude estimates, however, hide both the expansion of rain-fed agriculture over the last 150 years into areas of the global arid lands which had seen neither hoe nor plough before, and the fact that currently the Group I–III 'core' arid nations contain some 18 per cent of the world's cropped land.

The expansion of rain-fed agriculture into the arid lands in the last 100–150 years has taken two forms: first, the expansion of long-established systems already existing in, or very close to the fringe of, the arid lands and second, the expansion of European systems of mechanical farming into the arid lands of the Americas, southern Africa, central Asia, and Australia. In both cases the incentive was partly the pressure of an increasing population and the need for new food supplies. The main difference, established at least 100 years ago, was that the produce from the second (European) system was mainly destined for export back to Europe, whereas the traditional systems were producing primarily for local markets. It will be worth examining these two systems separately before we try to generalize on the question of how successful these attempts have been in spreading agriculture into lands climatically, economically, and socially 'marginal'.

There is no doubt that the expansion of agriculture into the arid lands has meant that currently a significant proportion of the world's basic food grains are grown in the arid lands (Table 12.1). The data are impressive, bearing in mind the crudity of the estimates and the possibility that the statistics underestimate actual production because of the difficulty of collecting data in the remote areas and because much of the production in the developing nations is consumed locally and therefore may not be noted officially. For both subtropical grains (millet and sorghum), together with the temperate

167

Table 12.1 Food production in arid nations

Arid nations	Products (1975–9 averages) (000 tonnes)				
	Wheat	*Millet*	*Rice*	*Maize*	*Sorghum*[a]
Group I 100% arid	2 055	810	2 384	3 039	227
Group II 75–99% arid	37 536	4 408	7 277	4 608	5 359
Group III 50–74% arid	29 260	562	584	19 281	7 184
Sub-total Groups I–III (% of world)	68 851 (16.9)	5 780 (15.3)	10 245 (2.8)	26 928 (7.6)	12 770 (18.6)
Group IV 25–49% arid	137 509	24 746	213 603	220 174	45 089
Group V 1–24% arid	116 012	5 985	16 249	40 365	5 033
Total arid nations (% of world)	322 372 (79.2)	36 511 (96.7)	240 097 (65.0)	287 467 (81.6)	62 892 (91.9)
World total	406 976	37 752	369 142	352 344	68 425

[a] 1977–9 averages only. *Sources*: FAO *Year Books*.

grain (wheat), production from the Group I–III arid nations seems to average 15–19 per cent of the global production. Not surprisingly, production of the grains having higher moisture requirements, namely maize and rice, is less impressive, being only 8 per cent and 3 per cent of global figures respectively. In addition, substantial amounts of wheat and sorghum are grown in the semi-arid areas of China, India, the USA, and the USSR, although the difficulty of separating out this production from that of the humid areas of these Group IV and V nations prevents detailed breakdown of the national figures.

Traditional systems

The origins of agriculture

Historians have debated the rationale behind the world-wide evidence for the domestication of plants by 8000 BC. Certainly by 7000–6500 BC archaeological evidence suggests that ancestors of most of the currently consumed grains were being cultivated for food to support an agricultural population which, in western Iran at least, was probably as numerous (27 km^{-2}) as, and better fed than, the present population (Braidwood 1971). Parallel situations seem to have occurred in central America, southeast Asia, and China and the debates on the rationale for agriculture have been briefly noted in Chapter 10. Although it meant more work for longer hours with less variety of food, agriculture at least provided more calories per unit area and this was the crucial factor (Cohen 1977). When combined with livestock grazing, it was even more attractive since animal protein was added to the diet.

Whichever explanation we accept, the fact remains that agriculture replaced hunting–gathering as an attempt to obtain a more intensive resource use. This use in the arid lands was extended into environments where, from hindsight, it had a limited chance of success. They were in fact, and have remained, the marginal lands for agriculture.

The system

The characteristics of rain-fed agriculture are the manipulation of soil to enable domesticated plants to bear maximum crops. Manipulation of the water inputs to the soil is minimal. In contrast to irrigated agriculture, the only 'manipulation' is indirect through the timing of the planting to coincide with optimal soil moisture conditions, the choice of plants able to tolerate the expected soil moisture conditions, and tillage practices which either try to conserve soil moisture prior to planting or reduce losses of soil moisture after planting, particularly from weeds. The system therefore implies the planting of seeds in a prepared seedbed and the protection of the plants from competitors or grazing animals until harvesting has been achieved. Such systems can be individually or group organized and, as we shall see, the labour inputs can and have been successfully mechanized.

Traditionally, cropping was continuous until the soil nutrients had been reduced sufficiently to make a significant impact on harvests, when the fields would be abandoned and new ground prepared as the seedbed (see Fig. 10.3). The basic temperate grain crops (wheat, rye, barley, and oats) as well as tropical grains such as millet and rice are some of the oldest cultivated plants. The refinements of the plant resources, as we saw in Chapter 6, have come from plant-breeding. The refinements of tillage practices have come mainly from mechanization.

Traditional systems in the 1980s

One of the largest areas of traditional agricultural systems impinging on the arid lands in the 1980s lies in the African continent north and south of the Sahara. In the south, in the area of the Sahel, stretching from the Atlantic coast to the shores of the Indian Ocean, some 25 million people practise a mixture of traditional shifting agriculture, limited irrigation, and livestock grazing. Kates (1981) identified three major 'livelihood systems'. First were rain-fed smallholdings producing food grains such as millet or sorghum alongside limited areas of cash crops such as cotton or peanuts, supporting

169

about two-thirds of the population. Cultivated areas had to be moved as soils became exhausted and, with increasing population pressure, return periods for particular plots had been reduced from 15 years in the 1930s to 5 years in the 1970s. A second major livelihood system (supporting 11% of population) was that of mixed agriculture and pastoral activities, where movement of livestock south to the humid fringe in the dry season and north towards the desert in the rainy season was used to provide manure in return for grazing on the stubbles of the grain crops. The third system was pastoralism where the sole or major livelihood for 13 per cent of the population came from their livestock alone. Again, a seasonal northerly shift to the edge of the desert in the rainy season to escape the tsetse-flies and return south when the rains failed was evident. The overall picture, except for the few islands of intensive, usually irrigated, cash crop production (often under foreign control), was of a complex but inherently traditional series of resource-management systems.

In the Sudan attempts to inject a new level of commercial mixed farming into traditional subsistence systems in the 1960s and 1970s seem to have failed because the planners did not allow for traditional 'satisficer' rather than 'optimizer' attitudes to resource use. Traditional farmers seemed unwilling to take risks with the new crops on the one hand, and on the other saw their future as part-time farmers with new part-time *urban* employment rather than as full-time commercially oriented farmers (Briggs 1978; Thimm 1979). This gap between the planners' and the local population's perception of the potentials for development together with ill-prepared and hasty implementation of the new schemes resulted in their failure.

On the northern edge of the Sahara there was more evidence of European-style agriculture, both in the new oases (Ch. 13) and in more successful government-sponsored mechanized dry-farming experiments, but the bulk of the rural population still practises traditional agriculture with wooden ploughs and animal power. Such systems included skilful 'opportunistic cropping' where local terrains concentrated the scanty run-off in moister hollows. In the Moroccan Sahara exposed limestone formations contained solution hollows (*dayas*) where run-off and blown soil accumulations supported periodic crops. With favourable rains, barley and wheat yields there could be up to two-thirds those in the oases, with the grain for human food and the stubble as grazing for livestock (Mitchell and Willimott 1974). Traditional systems could be remarkably efficient.

Agriculture in the arid land settlement process after 1800

From hindsight we can suggest models of how agriculture, particularly rain-fed agriculture, was used in the process of arid land settlement in the nineteenth and twentieth centuries. The bulk of the new agricultural settlements moved into land where the resource use had been previously less intensive, either hunting–gathering economies or pastoralism (nomadic or sedentary). As a result, not only was the human population density increased, but new plant species for the crops and a new level of environmental manipulation – cultivation of the soil – were introduced.

One model of the process of resettlement might be: first, the dispossession of any indigenous non-agricultural population. For the hunter–gatherers this was usually a physical ejection (American Indians, Aborigines) or a recruitment as labour for the new farmers (Hottentots, Bantus), while for nomads or ranching groups this might be by treaty or some form of financial compensation (American Indians, some Boer and Australian ranchers). A second stage would be the division by central government of the 'vacant' land into units for agricultural production and the allocation to potential farmers (as 'homesteads' in the USA, 'home maintenance' or 'living' areas in Australia). Given sufficient demand the area would be wholly occupied and, given successful farming operations, the next generation of farm-children would provide the subsequent demand for extra farms in not more than 15–20 years' time. This subsequent demand might be met by subdivision of existing units, if productivity and profitability had been increased sufficiently or if the population were prepared to take a reduction of living standards to stay on the land. If the demand could not be met, either the surplus population would have to look elsewhere for new farmland or move to other sectors of the economy, such as the cities, as a wage-labour force.

The variables which would affect the size and to some extent the fortunes of the agriculturalists are complex and interrelated. In Table 12.2 they have been separated into those which have tended in the past to work towards increasing or decreasing the size of the production units. As suggested in Table 12.2, the same variables might work both ways. The actual increase or decrease of unit sizes could be an arbitrary decision, perhaps as we shall see, made on political grounds.

The European system

From the mid-nineteenth century onwards the systems of agriculture developed in Europe and exported to the European colonies around the world have been moving into the global semi-arid and arid lands. The period of maximum activity varied around the world, reflecting in part local historical and political factors and the extent of popular pressure for new agricultural lands. From the 1860s to the 1900s was the period of maximum new agricultural settlement on the Great Plains of the USA and as late as the 1930s in Canada. In South Africa the period was from the 1860s to the 1930s, in Argentina from the 1880s to the 1920s, in Australia from the 1860s to the 1930s with some renewed activity in the 1950s–1960s, and in Russian central Asia from the 1890s to the 1920s and then again in the 1950s with the Virgin Lands campaign.

171

Table 12.2 Variables affecting size of agricultural production units

Variable	Tendency for unit size to ...	
	Increase	Decrease
1. Changes in natural environment	Decrease of production per area because of increased aridity or soil erosion or reduced soil fertility	Increase of production per area because of increased humidity or control of soil erosion.
2. Management	Increased livestock grazing needs. Reversion to fallow-crop rotation. Monoculture leading to reduced soil fertility. Capacity to delegate production processes to others: from owner-operator to sharecropper to subcontracting to others ('side-walk' or 'suitcase' farming)	Replacement of livestock grazing by more intensive land use. Continuous cropping. Build-up of soil fertility through fertilizer application. Increasing reliance on individual owner effort, i.e. reduction in available energy input
3. Energy available	Mechanization of production processes (land clearance, tillage, harvest)	Reversion to human labour
4. Capital	Fuel costs low or decreasing. Additional funds available (family source, partnerships, co-operatives, company operations)	Fuel costs increasing or high. Reduced funds
5. Markets	Decreasing prices and increasing prices. Profitability increasing and profits available to buy extra land.	Increasing prices? Profitability increasing and belief that more people could share the returns from area leading to pressures to subdivide existing properties

Although similar to traditional rain-fed agriculture, the European agricultural system of the mid-nineteenth century had several differences. While the crops included the traditional bread grains, there was virtually no use of millet because of climatic constraints, and maize and several beans indigenous to the Americas had been incorporated into the farming systems. Sheep, cattle, goats, and horses were incorporated into the system for their motive power, protein, fibre, and skins. The farmers themselves had possibly a stronger tradition of individual action, but also included tightly organized groups based upon ethnic, social, or religious affiliation which were prepared to move and settle as a unit. Such groups included religious refugees such as Mennonites, Mormons, and Hutterites, as well as social refugees such as communist and other co-operatively oriented groups. In addition there were the capitalists, farmers who had made fortunes in Europe or elsewhere in the 'new lands" and were seeking to extend their luck to the arid lands.

Coping with the environment

For most farmers, however, their occupation of the new farmlands involved considerable energy outputs, not only as physical labour on the farm, but also as costs to get the produce to market. A characteristic feature and one of the major reasons for the success of the European system was the extent to which these massive energy demands were met and these remote arid lands integrated into the global agricultural system.

On the farms the period 1860s–1930s was one of intense, and at times desperate, attempts at innovation aimed at reducing the human labour inputs and reducing production costs. The quality of the tools was improved. Steel replaced iron ploughs, lighter equipment replaced heavier. The human sower broadcasting the seed on to fertile or stony ground was replaced by mechanical seed-drills which not only planted the seed but, from the 1890s onwards, applied artificial fertilizer to the seedbed at the same time. Similar equipment uprooted weeds and, from the 1850s to the 1890s, successfully mechanized the laborious harvesting procedures. Draught animals were replaced first by steam power and then in the second decade of the twentieth century by the internal combustion engine (Pl. 12.1). The effect was to reduce the cost in time and money of agricultural operations. In the USA:

> In 1880 . . . 20 man-hours were needed to harvest an acre of wheatland. Between 1909 and 1916, this number was reduced to 12.7 man-hours, and between 1917 and 1921 – that is, with the advent of full mechanization – to 10.7 hours. The following decade cut the figure to 6.1 (1934–6).   (Gideon 1948, 162)

By the 1960s with self-propelled mechanical combine harvesters the time has been reduced to 0.25 hour.

The costs of clearing land for agriculture also were reduced. The steel

Plate 12.1 Farm scene, western Kansas, USA, photographed 1978. Family farm in Kit Carson County, with relatively modest capital equipment. From left to right: farmhouse in wind-break, row of five corn bins, harrow implement (chisel-type) with outer sections folded back for easier transport, large quonset machinery shed for protection in severe winters, low shed and feedlot for cattle with two feed bins on right.

'gang' ploughs and disc-harrows chopped up the tough sod of the semi-arid grasslands, and in Australia 'scrub-bashers' of logs or old steam boilers dragged by horse-teams and later tractors uprooted the semi-arid mallee scrub woodland which was then burned to prevent regeneration. 'Stumpjump' ploughs jigged through the ashes, their spring-loaded shares kicking up over roots and stones which would have snapped-off the traditional fixed ploughshares.

Even at the farmgate, however, the product was still remote from the market and the trip involved substantial energy outlays. Only by the provision of the railway networks, which from the 1870s onwards were beginning to extend into the semi-arid lands, was the commercial export of the grains possible and the cost of imports of vital supplies and equipment for the new farming lands kept down. In the USA and South America the railways were privately built, being paid for by grants of land to the railway company usually *pro rata* for length of track constructed. Thus the railway companies became at once both providers of transport and real-estate agents for the lands, the sale of which was essential for any profit for their speculative enterprise. In South Africa and Australia from the 1870s onwards, in Canada from the 1880s, and in the USSR after 1917 the railways were built by the central governments, as private capital was unwilling or unable to take up the challenge. In Australia the capitalists considered the land quality so poor that the size of land grants needed to satisfy them would have created unacceptable land monopolies which were rejected by the colonial governments.

By the construction of the railway networks, the detailed layout of which in Australia and Canada was keyed to the maximum distance wheat could be hauled to the rail siding (thought in the 1920s to be 24 km), and the 'tapering' of rail freight charges, the governments in these countries and in the USSR after the Revolution, were directly intervening in the land settlement process. This intervention was beyond the 'normal' process of surveying land into production units and supervising its allocation, but was aimed specifically at providing essential services – in this case communications.

By the time of the World Depression of the 1930s, which marked the end of the major period of agricultural advance into the arid lands, the bulk of the areas currently in agricultural production already had been developed. In these areas the surveyors' grids of roads and railways had been imposed upon the landscape. Individual (mainly monoculture) wheat farms were scattered at regular intervals over the surveyed blocks, with here and there a group settlement village in addition to the small service centres regularly spaced along the railway and main road alignments. Windmills raised domestic water from underground aquifers or surface run-off was collected in excavated tanks (especially in the Canadian prairies). In the service centres the blacksmith became in time the motor mechanic and garage proprietor. The bagged wheat shed alongside the railway station was transformed into the towering wooden or concrete silos. Shops appeared, churches and meeting halls were built, and communities came into being. For most of these areas this was the period of maximum population numbers. Soon afterwards the effects of the World Depression and the economic and technological adjustments produced by it were to begin to siphon off population and introduce some of the problems still facing such areas in the 1980s.

The exceptions to these generalizations were areas where agriculture continued to expand into new lands at later dates and where the peak of rural population densities associated with this intensification of resource use was delayed. Pockets of rangelands in the Great Plains were not ploughed up until the 1950s. Large areas of Western Australian sand-plains (for a brief period over 400 000 ha yr^{-1}) were ploughed up for wheat-farming in the 1960s. The massive Soviet plough-up of 32 million ha of semi-arid steppes (the Virgin lands) in Kazakhstan and Western Siberia took place over the period 1954–7. In Chinese Turkestan new collective farms were established after 1949 and again after the arrival of the railway in 1959 (Wiens 1969).

In the USA and Australia individual farmers and their farms moved in with government support in the provision of services, some financial help, official land appraisals, and agricultural advice (particularly on soil nutrient status and tillage methods). Thus, in the case of Western Australia, wheat-farming was and still is successfully carried out with an average rainfall of only 200 mm. In Soviet central Asia the nucleated State farm village settlements surrounded by their ploughed fields were pushed beyond the 250 mm to the 150 mm annual rainfall isohyet – as in Western Australia on sandy soils with good moisture-retention qualities.

Yet by 1980 there was evidence in many of these agricultural areas of environmental stresses and a decaying social structure alongside evidence of considerable individual wealth. In the Americas and Australia derelict and abandoned farmhouses were interspersed with obviously successful farmsteads across the countrysides, empty shops could be found in the service centres, and children had to travel long distances often by public bus even to primary school. The fields still produced the grains although there were more areas of sown grasses and patches of eroded land or salt-impregnated low spots broke the even pattern of productive land. The grain silos were generally full, although drought could still empty them from one year to the next. The laden railway trucks still trundled the grain to the ocean cargo terminals, but along fewer routes than before. What had happened to produce this complex picture of success alongside obvious failure? To answer this question we need to ask another. What *is* successful land settlement and has it been achieved on these semi-arid agricultural lands? Some answers can be provided from the Americas and Australia.

Success or failure?

Success in land settlement might be measured by a variety of criteria. Some possibilities are as follows:

1. Has production been increased and has that increase been maintained through periods of environmental stress, e.g. droughts?
2. Has the resource use system increased and maintained the human population-carrying capacity of the area?
3. Has the quality of the environment been maintained or improved in the area?
4. Has the settlement been self-sufficient or has it required subsidies from elsewhere, e.g. national finances?

We shall examine the historical geography of rain-fed agriculture in the North American and Australian arid lands in the light of these questions.

Agricultural production trends

The invasion of the Great Plains and Australian arid lands by farmers from the 1860s onwards brought a rapid increase in productivity. Not only were grain crops produced where none had been before, but they were produced in significant quantities and for a brief period yields per area also increased. Yet droughts on the Great Plains in the 1870s, 1880s, and 1890s and in southern Australia in the 1880s greatly reduced yields. Many farmers were bankrupted and left their properties. Disaster relief had to be organized for the survivors. This relief varied from basic food, water, and clothing to free replacement seed for the next planting.

Further failures of the wheat harvests were noted in the droughts of the 1930s in both the Great Plains and Australia. Over the period 1931–54, studies of wheat failure on the Great Plains showed that for over half the years drought was the main cause of failure, with hail, grasshoppers, and soil erosion also important factors. Losses were up to between 60 per cent and 80 per cent of the planted acreage (Hewes 1965). In northwest Kansas over the period 1915–48 the fortunes of the farmers varied (Table 12.3). Whether the fortunes reflected drought or other environmental stresses, market price fluctuations (the period included the Depression), or bad management, was not indicated but 'the need for credit [or a change in resource use] is obvious' (Hewes and Schmieding 1956).

Table 12.3 Economic Stress in the Great Plains rain-fed agriculture 1915–48

Farmers' economic position at beginning of period	Years when lost money	Years when did not meet living expenses	
	No. (%)	No. (%)	Longest run of such years
1. Debt-free owner-occupier	3 (9%)	10 (30%)	5
2. Tenant	0	11 (33%)	7
3. Debt-encumbered owner-occupier	7 (21%)	14 (42%)	8

Note: Data are for northwest Kansas.
Source: Hewes and Schmieding 1956.

In the Mallee area of southern Australia where wheat farms expanded rapidly after the First World War, production was hit by droughts in 1914–16, in the early 1930s, in 1945–6, in 1957–9, and in 1965–6 (Fig. 12.1). The 1930s saw several years with costs greater than return. Even the significant improvement in yields from the 1950s onwards (as the result of the introduction of crop rotations with nitrogen-fixing legumes such as *Medicago* spp., and greater use of phosphate fertilizers) could still show the effects of droughts (Heathcote 1980; Williams 1974).

The answer to the question on productivity seems to be that production has been increased, although the impact of droughts can still bring substantial reductions in harvests with associated economic losses sufficient to cause substantial hardship.

Population carrying capacities
Examination of census data shows that for the Great Plains most of the rural areas reached a peak of population numbers and densities in the early 1930s

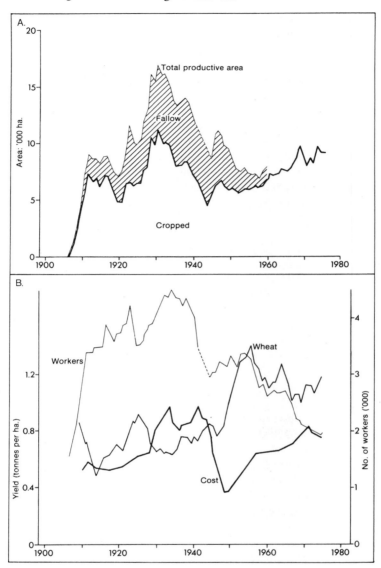

Fig. 12.1 Rain-fed agriculture in semi-arid Australia, trends 1900–1970s. *Source*:
Fawcett 1978
(A) *The rise and fall of dry-farming using fallowing*
Notes: Although cropped land in the late 1970s had almost returned to
1930s levels, the land fallowed as part of the cropping system had declined
to insignificant proportions. Past fallow land was now in rotation pastures.
Data are for County Chandos in the Murray-Mallee of South Australia
(B) *The changing workforce and economic viability of wheat production in
the Murray-Mallee*

and generally have lost population since then. Borchert (1971) noted that since the 1940s the Great Plains had been the largest continuous area of rural population decline in the USA. In the Australian Mallee the rural workforce declined from the peak in the 1930s to about half the numbers by the 1970s with increased mechanization and increased farm size. This decline was forecast to continue into the 1980s as farmers expanded their properties to achieve economies of scale (Fig. 12.1).

Associated with the run-down of rural population from the 1930s was a reduction of population in the smaller regional centres of the Great Plains and the Mallee. In the 1970s the urban centres of 1000–2500 population benefited from industrial development in the USA but the Australian towns continued to lose approximately 10 per cent of their population every 5 years. Some of the reasons suggested were the independent local government system and the possibility of lower rural wages in the USA, compared with the lack of effective local government and standard rural–urban wage rates in Australia (Lonsdale and Holmes 1981).

The run-down of rural population was in some cases paralleled by an increase of absentee owners, who might also have other sources of income. Kollmorgen's studies of these absentees on the Great Plains in the 1950s identified both 'sidewalk' and 'suitcase' farmers – the sidewalk farmers living not on their land but in the nearest town, while the suitcase farmers lived more than 48 km away from the border of the county in which their land lay. In South Dakota they held 8 per cent and 15 per cent respectively of the land, but 36 per cent and 70 per cent of the land in cash crops. In some Kansas counties they could comprise up to half the total owners. From 2100 km away in Florida some farmers controlled the resource use by means of sub-contracting – often of the whole process, from seedbed preparation to harvest (Kollmorgen and Jenks 1958). Similar kinds of town-farming have been identified in Australia (Williams 1974).

The question of population carrying capacity, therefore, must similarly be given a qualified answer. The carrying capacity has increased overall, but currently is declining and the decline will continue. Associated with this decline has come greater remote control of resource use by managers who are absent most of the time, have lost the resident farmer's intimate contact with the land, and who therefore would be unable to keep close touch with environmental fluctuations on their properties.

Notes: The rural workforce peaked in the 1930s, but had been more than halved by the 1970s.

The cost curve indicates the average wheat yield (5-year running means) required to cover costs of production. Actual average wheat yields (5-year running means) peaked in the 1950s, providing thereafter a more stable source of income than the previous 40 years.

Data are for County Chandos except for the workforce which represents data for four counties including Chandos.

The environmental impact

The widespread plough-up of the semi-arid grassland and shrublands – placing the sod in the famous comment of a chief of the Sioux tribe 'wrong side up' (Mollin 1938) – led to massive soil erosion around the world. The potential for erosion was further increased after 1900 by the widespread adoption of 'dry-farming' methods to try to cope with the drought problem.

Devised by an American farmer, H. W. Campbell, who published his first descriptive pamphlet in 1893 and a series of technical manuals between 1902 and 1916, the dry-farming system required deep ploughing to break up the soil to receive and hold any precipitation; subsurface packing to prevent seepage of moisture down the profile out of reach of the plant roots; and, after precipitation had been received, constant maintenance of a loose soil surface mulch to reduce loss of moisture by evaporation. The intention was to carry out this process on fallowed land for one year and then plant in the second year, thus in theory providing two seasons' rainfall for the crop (Hargreaves 1957). Subsequent research has shown that fallowing with dry-farming methods only conserves about a quarter of the previous year's rainfall on the Great Plains, and only about 10 per cent in the Mallee. At the time, however, the technique was widely publicized by the US railway companies (as a means to sell their lands, the drought potential of which had already become obvious) and accepted by Great Plains farmers. In 1906 the South Australian Surveyor-General visited Campbell's operations and returned so impressed that he persuaded the State government to advocate its use on the drier wheat lands.

By the 1930s the effects of the widespread fallowing and continuous tillage had created a dangerously loose soil texture. The combination of droughts, high winds, and the frantic plough-up of large areas to increase crop returns at times of falling wheat prices led to widespread dust storms. From the 'Dust Bowl' of the southern Great Plains in 1935 dust fell on the capital Washington. Similarly, in southern Australia dust from the eroding wheatfields coloured the sunsets of the eastern capitals and in the 1940s even reached the New Zealand snow-fields. Crops were blown away, roads and railways blocked, irrigation canals filled in, aircraft flights grounded, buildings buried, and people affected by dust inhalation. These major 'natural disasters' brought massive disaster relief from the state and national governments. Subsequently, above-average rainfalls and guaranteed prices for produce during the Second World War helped the rehabilitation of the eroded lands.

By the 1980s, evidence had accumulated both of improvement in some aspects of the soil erosion problem and the appearance of a new threat to the farmers from increasing soil salinity. Comparing the 1980s with the 1930s there are larger areas of the semi-arid and arid areas now protected against development for agriculture, either by reserves of the remaining natural vegetation as national parks or wildlife reserves or by reversion to grasslands for controlled ranching operations as in the various 'national grasslands' in

Plate 12.2 Strip-farming on the Great Plains, USA, photographed 1968. The alternating strips of ploughed land (dark tone) and natural or sown pasture
(light tone) are at right angles to the prevailing westerly winds. The shallow valleys of the eastward-falling watercourses have been left as natural
grazing (light tone). Photographed from about 7000 m, view to northwest.
The same area in early 1981 showed a less dominant but still obvious
pattern of strip-cropping with considerable areas under centre-pivot
irrigation.

the USA. Further, in the cultivated areas there is evidence of the wider use
of soil-conservation techniques such as strip-cropping, stubble mulching
(leaving the crop stubble residues as protection of the soil surface until the
seedbed for the next crop is prepared) and minimum tillage (using chemical
sprays rather than mechanical tillage to control weeds, Pl. 12.2). The results
have been reduced levels of soil erosion since the 1930s, although droughts
in the 1950s and 1970s still brought dust clouds to the Great Plains
(Fig. 12.2) and the problem of erosion has not been solved in Australia
(Table 12.4).

On the negative side a new threat in the form of increasing soil salinity
has appeared. Summer fallowing of croplands on the northern Great Plains
apparently had allowed ground-water levels to build up over 20 years since
its introduction on heavy glacial till soils over impervious substrate. As a
result, by the 1950s reports of soil salinity build-up affecting crop yields were
appearing. Current estimates suggest that these areas, which cannot be further cultivated without costly remedial measures, are expanding at 2–10 per
cent per year (Heathcote, 1980). In the Mallee of southern Australia and
the sand plains of Western Australia clearance of deep-rooted perennial vegetation and replacement by shallow-rooted grain crops has caused similar

181

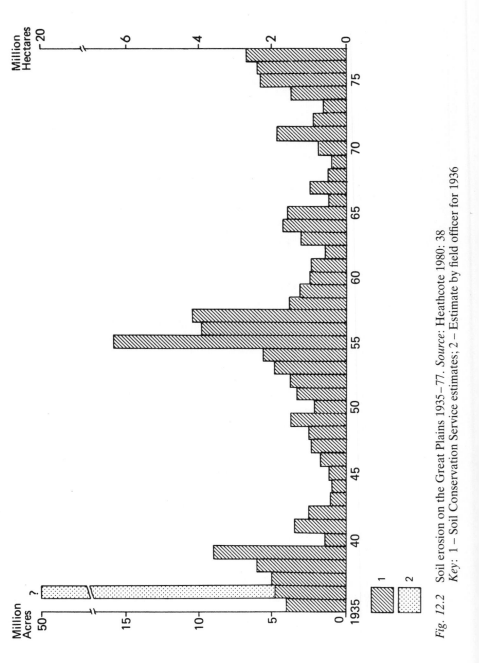

Fig. 12.2 Soil erosion on the Great Plains 1935–77. *Source:* Heathcote 1980: 38
Key: 1 – Soil Conservation Service estimates; 2 – Estimate by field officer for 1936

Table 12.4 Soil conservation needs in the rain-fed agricultural areas of Australia

Soil conservation measures	Extensive cropping areas[a] %
Not required	31.6
Management only	31.6
Works needed	33.6
Works already applied	3.1
Change of land use needed	0.1
Total area in use (000 km^2)	443 (100%)

[a] These include arid and humid areas, but the data do not allow for distinction between them.
Source: Australia 1978.

rises in ground-water levels as less water is lost by evapotranspiration, with resultant salting of soils and loss of crop area (Mabbutt 1978). Paradoxically, in these semi-arid areas the effects of annual cropping have been to *increase* the amounts and levels of ground-water to a point where excess water in the soil is a problem at the same time that seasonal shortages of rainfall inputs may occur. A better example of tragic ignorance of the ecology of the areas could hardly be found.

Like the others, the question of environmental quality therefore has to be answered guardedly. There has been massive environmental deterioration from soil erosion and soil salinization, but some of those effects have been offset by the inherent fertility of some of the deep Great Plain loess soils, massive inputs of chemical fertilizers elsewhere, and by soil-conservation measures. The erosion threat remains, however, and since several of the conservative measures reduce the potential crop area (as wind-breaks, strips of stubble left unused for a year, and terracing of fields) they are not particularly attractive to farmers anxious to make money quickly. Indeed one of the problems of the semi-arid grain-farming lands has been that given the right sequence of above-average rainfall seasons the lucky speculator can, with one or two years' harvests, make a substantial profit, when a few poor seasons earlier or later he would have been bankrupted. For those farmers willing to take risks the semi-arid fringe of rain-fed agriculture has a continuing fascination, but we must not forget that part of the speculators' success has come from the willingness of national governments to subsidize such settlement by disaster relief in times of economic or environmental stress.

Self-sufficient or subsidized settlement?
In the complex regional interchanges and flow of goods and services in a modern state it is increasingly difficult to say that particular regions are donors or debtors to the national treasury. Wool bales from the remote arid pastoral stations, for example, could be said to provide employment for

stock and station agents, railway and dock workers, jute growers in Bangladesh, urban textile mills, and ships' crews, as well as benefits to the pastoralists themselves. In some cases however, it is possible to identify specific transactions which transfer national funds to regional areas for specific purposes, either as relief and rehabilitation after some natural disaster – to re-establish the *status quo ante*, or as development funds to finance a change of resource use which it is hoped will increase the contribution of the region to the national welfare in the future. The semi-arid rain-fed agricultural areas have benefited from national transfers of both the above types and thus for certain periods of time, because of their inability to support their population from their own resources, became subsidized settlements.

The Great Plains as a whole has been one of the largest areas in the USA to receive federal funding in various forms. Studies by Borchert (1971) and Hewes (1973) have shown that massive national assistance (some $375 million alone requested in 1934–5) was provided to the Dust Bowl area during the crisis years of the 1930s. This was relief for the immediate distress caused by drought, low prices for produce, and the devastation by soil erosion. The relief took the form of loans, free petrol for farmers' tractors to carry out soil-conservation measures, and local public works jobs to provide income. In addition the relief tried to change what were recognized to be disastrous resource-management techniques, by encouraging farmers to sow grass instead of grains, and by purchasing the worst eroded areas for treatment and recuperation as 'national grasslands'. In the 1950s, at a time of agricultural surpluses, the farmers were paid to take their land out of crop production for 10 years and sow it down to grass, under the Soil Bank Scheme (1956–60). The argument here was that the land under grass would be still producing feed for livestock, but would not be providing further *crop* surpluses. The land would be protected from soil erosion, and would be available, if required in the future, for plough up again as cropland. By 1960 some 11.6 million ha, mostly in the old Dust Bowl area, were under grass in the scheme. The national resources – in this case the soil – had been 'banked' to solve an immediate problem of oversupply and to provide a resource for the future.

A similar series of state and national programmes were applied to the semi-arid farming lands in southeastern Australia. The economic depression of the 1930s, the demonstrated failure of several of the soldier settlement schemes on the drier fringes of the farmlands, together with droughts in the 1930s and early 1940s, brought farm bankrupties and foreclosures, population outmigration, and widespread soil erosion. Official drought-relief schemes – free replacement seed, free freight for starving livestock, and public works jobs – became more frequent and eventually were merged into the 'Marginal Lands Scheme' (1939–61) by which $4 million of national funds was allocated to allow the lands of bankrupt farmers to be purchased. These were reallocated to the surviving neighbours who, on the larger areas, would be required to change their land use from monoculture wheat to mixed crop and livestock with crop–pasture rotation to reduce the erosion hazards (Heathcote 1980). The results, as on the Great Plains, were improved farming methods but at the cost of a halving the number of farmers and considerable national funds.

A further source of regional subsidy has been the federal crop insurance schemes peculiar to the USA. Schemes for private insurance against hail damage on the Great Plains date from at least 1883, but in 1919 several states introduced their own schemes and after some experiments in 1938 the federal government offered insurance to wheat farmers for drought and frost as well as hail. By the mid-1970s the Great Plains area as a whole was receiving 47 per cent of the federal crop insurance indemnity payments and 46 per cent of the private company insurance payments. Over the period 1948–55 the counties at the heart of the 1930s Dust Bowl had amassed a deficit of over \$18 million (indemnities less premiums) with the federal scheme, and in southeastern Colorado the losses were so great that the scheme was withdrawn (Heathcote 1980).

For a brief period in the 1960s, the bedouin nomads being resettled as farmers in the Negev by the Israeli Government had a profitable existence through what became known as 'compensation farming' (Amiran and Ben-Arieh 1963). They planted crops in any possible location. If the rains fell they reaped the harvest and sold at a profit. If the rains did not fall they claimed and got from the government compensation for their 'lost' crops. The same principle was alive and well on the Great Plains in the 1970s.

In terms of the four questions posed about recent farming in the semi-arid lands, the success of settlement has been shown by the rapid increase in the intensification of resource use, production per unit area, and population to the 1930s. Thereafter the exploitative nature of that resource use has been somewhat curbed. The areas are still zones of high-risk farming and marginally successful settlements. Political decisions to maintain settlement in the areas have carried high costs both to the environment and the national treasuries.

CHAPTER 13

Irrigation: panacea or Pandora's box?

Irrigation, the 'artificial application of water to land for agricultural purposes' (Gulhati 1955: i), has been one of the oldest adaptations of resource management in the arid lands. Although covering a relatively small area of the arid nations, some 3.7 per cent in 1978 (Table 13.1), the intensification of resource use and outputs that irrigation allows through the boosting of available water supplies, means that the production and population supported by the irrigated lands is out of all proportion to the area actually irrigated. If Egypt is the gift of the Nile it is because the waters from the Nile gave Egypt an effective rainfall of 2083 mm instead of the actual average of 127 mm, plus the rich nutrients in the flood sediments. The devel-

Table 13.1 Irrigation in the arid nations 1961–78

Nations	Area irrigated (000 ha)		Percentage increase
	1961–5	*1978*	*1961–5 –1978*
Group I	3 010	3 477	+ 15.5%
Group II	23 270	29 573	+ 27.1%
Group III	3 285	4 672	+ 42.3%
Subtotal			
Groups I–III	29 563	37 722	+ 27.5%
(% of world)	(19.8)	(18.8)	
Group IV	82 775	106 712	+ 28.9%
Group V	12 771	21 051	+ 64.8%
Total arid nations	125 109	165 485	+ 32.3%
(% of arid nation area)	(2.8)	(3.7)	
(% of world's irrigated area)	(83.7)	(82.4)	
World total	149 474	200 913	+ 34.4%

Source: FAO *Production Year Books.*

186

opment and application of irrigation technology in and beyond Egypt is therefore testimony to the skilful hands and imaginative brains which have diverted these natural 'gifts' to benefit mankind. However, the problems that technology has created are testimony to dangers of massive manipulation of the environments without adequate knowledge of the countless effects of those manipulations.

Global irrigation patterns

A survey in 1955 and data from 1965 provide some basic facts on global irrigation patterns (Table 13.2). In 1955 the top twenty countries in terms of area irrigated had 91 per cent of the world's irrigated lands. By 1965, depending upon which source was used, this was still either 90 per cent or 93 per cent. The top five countries had 66 per cent of the total in 1955 and either 72 per cent or 74 per cent by 1965. In terms of national units, irrigation is concentrated in few countries.

The data shows that irrigation is not confined to the arid lands, although arid nations dominate the top ranks at both dates. In humid nations such as Indonesia, Japan, and the Netherlands, as well as the non-arid parts of China, irrigation is used as a guarantee of crop yields or as a means of intensifying cropping patterns over time.

The national role of irrigation varies considerably between countries. In Egypt the small area of cultivated land is all irrigated. In Argentina the figure in 1955 was only 3 per cent. There were great variations between these extremes. The largest proportion of the total national area irrigated in 1955 was not in the arid lands at all, but in the Netherlands. This pattern reflected the use of irrigation on the polder lands and intensive horticultural areas as a guarantee of controlled soil moisture conditions. More usually, however, the irrigated areas represent areas less than 10 per cent and often less than 5 per cent of the national land areas. Yet, as we shall see, the relative size of these areas is often inversely related to the relative importance of their production.

The area under irrigation on a global scale has been expanding rapidly over the last 70 years and that rate of expansion has been increasing over time (Table 13.2A). From 1900 to 1930 the area almost doubled. Subsequently, the rate of increase doubled and by the 1970s almost 4 million additional hectares a year were being irrigated. What is not clear from these figures is the extent of old irrigated lands which have been lost at the same time, since the global figures are lands actually irrigated. However, there is no doubt of the size of the expansion of irrigation which is occurring.

The irrigated area in the arid nations expanded by a third from the 1960s to 1978 (Table 13.1), although the share of the global irrigated area dropped fractionally. Nevertheless, the arid nations still contained over 80 per cent

187

Table 13.2 Global irrigation areas 1900–78
A. Expansion of global irrigation areas

Date	1900	1930	1955	1965	1968	1978
Area (million ha)	44[a]	80[b]	120[c]	162[d] or 172.8[e]	162.6[f]	200.9[f]
Annual rate (million ha)		1.2		1.6	3.2	3.8
(Dates)		(1900–30)	(1930–55)	(1955–68)		(1968–78)

B. Top twenty irrigated nations

Country	Area (million ha)	1955[e] Percentages of area			Areas (million ha)		
		Cultivated	Irrigated	Cultivated actually irrigated	1965[d]	1968[f]	1978[f]
1. China	30.9	13.93	3.21	22.98	52.8	41.0	48.0
2. India	23.6	36.56	7.28	19.92	25.6	27.2	35.5
3. USA	10.5	17.63	1.36	7.69	15.2	15.7	16.7
4. Pakistan	8.5	19.23	9.11	47.35	10.8	13.1	14.0
5. USSR	6.4	?	0.29	?	12.0	10.2	16.6
6. Indonesia	4.2	5.78	2.27	39.23	3.6	4.2	5.3
7. Japan	3.8	13.71	1.03	75.57	3.2	3.3	3.3
8. Iraq	3.2	?	7.42	?	3.6	1.4	1.7
9. Egypt	2.8	2.83	2.83	100.00	2.8	2.8	2.8
10. France	2.4	57.38	4.53	7.90	2.4	0.7	0.8
11. Mexico	2.1	11.85	1.09	9.24	2.8	3.4	5.0
12. Italy	2.0	71.83	6.80	9.46		2.5	2.9
13. Iran	2.0	?	1.23	?	2.4	5.1	5.9
14. Afghanistan	1.6	4.17	2.67	64.00		2.3	2.6
15. Peru	1.3	12.78	1.04	8.13	1.2	1.1	1.2
16. Chile	1.3	8.49	1.73	20.39	1.2	1.1	1.3
17. Netherlands	1.0	?	25.00	?		0.2	0.3
18. Argentina	1.0	10.92	0.36	3.33	1.2	1.2	1.5
19. Thailand	0.8	10.56	1.72	16.30		1.8	2.6
20. Australia	0.6	1.10	0.08	7.95	0.8	1.4	1.5

Sources: [a] Encyclopaedia Britannica; [b] Carrier 1932; [c] Gulhati 1955; [d] Int. Commission on Irrigation and Drainage; [e] FAO estimate; [f] FAO *Year Books.*

of the global irrigated lands and the core and predominantly arid nations still retained about a quarter of the global figures.

The major increases in irrigated area seem to have occurred in the semi-arid (Group IV) and peripherally arid (Group V) nations. In the latter case the increase over the period was 65 per cent, which included increases in the

USSR of 73 per cent and in Brazil of 92 per cent. The extent to which such increases are taking place within the arid zone or are being used to complement humid resource-use systems remains to be seen.

Irrigation systems

Since a variety of methods are used to bring water to the fields by artificial means and since those methods are related to different field site and water source characteristics, we need to describe the variations in those characteristics to assess the resource-management potentials.

Field sites and water sources

A basic distinction needs to be made between surface and subterranean water supplies since the capacities of related distribution systems vary between them.

Surface-water sources
Two basic types of field sites usually are associated with use of surface-water flows where a natural 'head' of water is directed to the fields using gravity as the energy supply. The world's floodplains provide naturally irrigated areas. Where the main stream flowed at higher levels than the surrounding valley floor the natural 'camber', resulting from silt accumulations closest to the river over time, enabled a simple lifting device or a breach in the natural levee banks at flood times to direct river water away from the watercourse on to the surrounding valley floor. This water contained silts and the deposits provided plant nutrients as well as gradually raising the level of the fields. Before the control of the Nile, deposits averaging 1 mm resulted from the annual flood and estimates put the soil accretion at 2.4 m over the last 1300 years (Hamdan 1961).

The pediment or bajada sites similarly offered natural floods. In these situations, steeper gradients meant faster and potentially more destructive water flows, particularly since there might be less warning of the floods. However, storage of waters along the watercourses would be easier than on the floodplains because of the smaller volumes of flow, narrower valleys and watercourses, and greater chance of solid rock at or close to the surface for dam foundations. Soils would be more varied in texture and might include coarse sands or gravels as part of the alluvial fans, and the areas controlled from a single water source would be less than on the floodplains.

Underground sources
Underground water supplies offer smaller volumes of water for immediate usage but, depending on the lifting mechanisms, more regular supplies. Unless the supplies are artesian, the deeper the aquifer the fewer and more costly the wells. A shallow aquifer which could be tapped by individually dug

and worked wells would result in a differently organized irrigation system than one where a very deep aquifer could be tapped only by a very expensive bore, from which water would have to be carefully allocated between users.

Irrigation techniques

Two major techniques of artificial application of water to the fields can be identified. One diverts natural gravity flows out of watercourses on to the fields. The other reticulates water over horizontal or vertical distances using other energy sources (Table 13.3).

Diversion techniques

The diversion techniques use naturally occurring peak water supplies as and when they occur; make no attempt to store those surface waters except in the soil; use both the water and the silt it carries on the fields; have only partial control of the amounts received; and, have control only of the direction of water flow, not of its timing.

Perhaps the oldest such system is the diversion weir or *seil* (Arabic for 'flood') where natural floodwaters coming down a wadi are diverted out of the watercourse on to the adjacent fields. One of the most famous of these weirs was that at Marib in the Yemen which was built about 600 BC to divert floods through sluices and guiding walls to garden plots on either side of the Wadi Adhana (Fig. 13.1A). The collapse of the structure in the sixth century AD is supposed to have caused the end of the civilization it supported. Such systems are still in use and represent approximately a quarter of the irrigated areas of the Wadis Hassan and Bana in South Yemen.

Equally ancient is the basin system developed first along the Nile banks when the regular seasonal floods were channelled into a series of rectangular basins (800–24 000 ha) bordered by dikes from which excess water spilled into the next basin downslope. Because of the size of the area and volume of the floodwater, communal efforts were needed to construct and maintain the dikes and divert the waters. The Nile, Tigris–Euphrates, and Hwang Ho valleys were among the largest areas where this system was applied.

Archaeological discoveries of the last 30 years have revealed a further system of 'run-off' farming in southwest Asia. Originally discovered in the Negev, the Nabataean system (named after the culture using it in the sixth century AD) required the careful guidance of surface run-off from a large 'catchment' area on to a small 'run-on' area which was sown with crops. A sequence of such run-on areas might form a flight of terraced fields on a wadi floor, to each of which run-off from a particular area of the surrounding uplands was guided by low walls and channels on the hillsides. Because of the high ratios of catchment (run-off) to usable (run-on) area, from 12 : 1 to 30 : 1, and risk of flood damage in the wadi floor the system needed a large area and was hazardous. Nonetheless it has been successfully demonstrated again in the central Negev (Evenari, Shanan, and Tadmor 1971).

Table 13.3 Irrigation systems: advantages and disadvantages

Criteria	Systems						
	Diversion		*Reticulation*				
	Seil	*Basin*	*Canal*	*Qanat*	*Well/tank*	*Spray*	*Drip*
Sites	Alluvial fans, bajadas	Floodplain	Fans to floodplains	Fans, bajadas	Fans, floodplain, plains	Plains	All are possible
Size of usual operating units	Single to group/region?	Regional	Single to national	Single to group	Single to group	Single	Single
Water supply characteristics	Seasonal Silt content Surface	Seasonal Silt content Surface	All year? No silt? Surface	All year No silt Subterranean	All year No silt Subterranean	All year Pumped No silt	All year Pumped No silt
Technology	Weirs and diversion walls	Dikes and sluices	Channels, dams, and sluices	Tunnels	Boring, pumping	Pumping, distribution systems	Pumping, distribution systems
Problems	Timing of water application Volume of water?	Drainage Timing and volume?	Maintenance Seepage Evaporation Silt traps	Capacity limited Maintenance	Lift costs Maintenance	Salting and 'burning' of leaves	Maintenance of nozzles

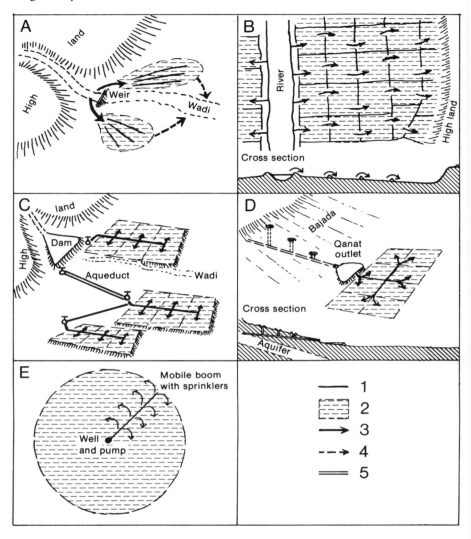

Fig. 13.1 Irrigation systems. *Sources*: See text
Key: (A) Seil (flood) system; (B) Basin system; (C) Reticulation system (dam and canals); (D) Reticulation system (*qanat* or *foggara*); (E) Centre pivot sprinkler system 1 – Dikes or walls; 2 – Irrigated areas; 3 – Water movement; 4 – Drainage of excess water; 5 – Canals

Reticulation techniques

The reticulation techniques attempt complete control of water supply and application to the field. Natural surpluses of surface water are stored in dams and underground supplies are tapped at will. Because the technique implies that water may be stored for periods of time, any silts carried are usually deposited in the storage dams or canals (Pl. 13.1). Most plant nutrients are thereby lost from the irrigation water. This silt further reduces both storage and transport capacities. Its removal is a regular maintenance cost of the technique.

The technology associated with the reticulation techniques has always been more complex than that in the diversion systems. Some of the human- and animal-powered lifting devices have been long common to both (Fig. 13.2), but the development of mathematics and engineering skills have been long associated with the irrigated civilizations of southwest Asia and China. Canal and *qanat* construction and field levelling to reduce the velocity of surface-water flows, increase percolation, and reduce soil erosion, required precise land and slope measurements. In the 1980s, for example, the most sophisticated irrigated field levelling in the USA uses laser beam technology (Pl. 13.2). The development of new techniques for dam construction has increased the capacities such that for example the entire annual flood of the Nile can be contained in the High Dam. The engineering problems and costs of construction in the 1980s, however, are still formidable. Failures still occur (e.g. the 1979 collapse of the Macchu II Dam in India

Plate 13.1 Gillespie Dam on Salt River, Arizona, USA, photographed 1979.
The view is downstream from a spur in the foreground overlooking the sediment-filled dam in the centre, whose parapet is virtually invisible but is parallel to, and just in front of, the piers and spans of the road bridge and towers supporting a gas pipeline across the river. The dam was completed in 1920 as part of the Salt River Project. Apart from the obvious silt problems, irrigators have had to face naturally high salinity in the river as its name indicates.

193

System	Lift (m)	Energy	Area irrigated per day (ha)
1. Archimedes Screw 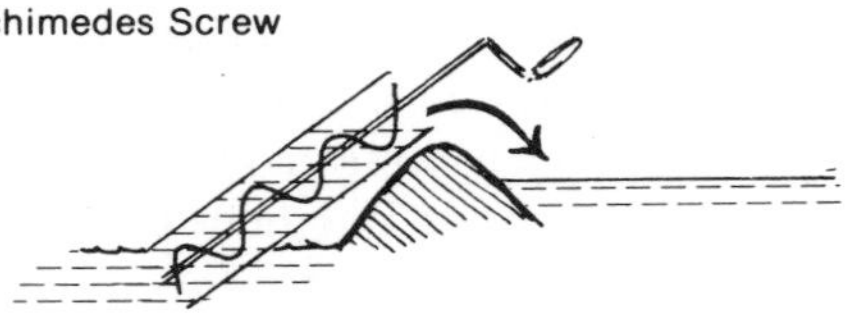	1	1 man	0.3
2. Shadouf 	2–5	1 man	0.1–0.3
3. Noria 	c.20	Water power	>0.8
4. Sakiya 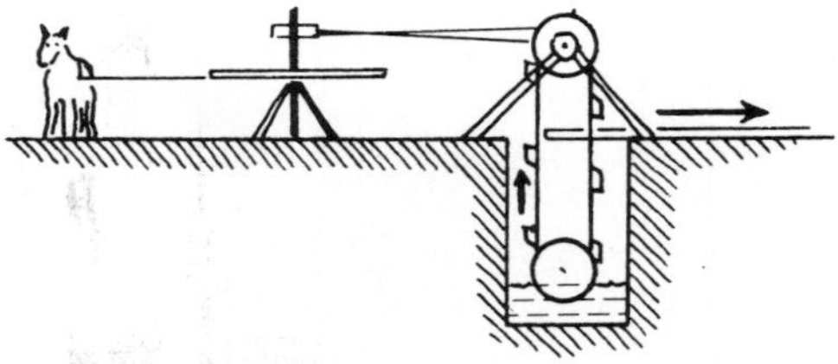	100?	Animal power	2–4.8

Fig. 13.2 Traditional water-lifting systems. *Sources*: Drower 1956; Stamp 1961 and various
For each system the lift in metres, energy source, and approximate area irrigated per day in hectares is indicated

Plate 13.2 Laser levelling of irrigation areas, Arizona, USA, photographed 1979.
The two earth-moving machines (1) are controlled by means of a rotating horizontal laser beam projected from the central tripod (2) with receivers on the machines adjusting the blades of the scrapers according to height above or below the beam. With this method the original surface height variations of up to 9 cm in a 4 ha field could be reduced to approximately 1.5 cm, with consequent improvement in watering efficiency. The site is on the Wellton–Mohawk Irrigation and Drainage Project, central western Arizona.
The white markers in the foreground show the bank left between the irrigation fields or 'basins'.

with the loss of 5000 lives). Increasing criticisms are being levelled at the social and environmental effects of the 'big dam' solutions to local food production problems in the arid lands.

Origins and rationale for irrigation

Origins

Irrigation systems appear to have developed independently in at least three major global locations; southwest Asia (including the Nile and Tigris–Euphrates valleys), China, and the Americas (Peru). The age of the earliest systems is not clear. In southwest Asia, sites in Iran are tentatively dated to 5500 BC. Basin systems were in operation along the Nile by 3000 BC. Canals in Mesopotamia were mentioned in the Code of Hammurabi of

2300 BC, have been dated to 1200 BC in Peru and at least 300 BC in China and India.

The seeding of fields or swamps drying out after natural flooding is suggested as the predecessor to man-induced flooding or irrigation – a kind of 'geological opportunism' (Vita-Finzi 1971). In the Euphrates Valley the natural camber of the floodplain allowed basin systems to use water up to 5 km away from the main watercourse (Flannery 1971). In cultivating these irrigated areas the population was concentrating its efforts on a smaller proportion of the land area. In prehistoric Iran the hunting–gathering folk were able to use about 35 per cent of the land surface for the most productive resource use; rain-fed agriculture enabled only 10 per cent of the area to be so used. When irrigation was introduced the percentage dropped to about 1 per cent. Yet the production from this reduced area under irrigation supported a population density around 3000 BC of up to 6 persons km^{-2} compared with 0.1 under hunting–gathering and 1.2 for rain-fed agriculture (Flannery 1971). This was but one of the attractions of irrigation as a resource use.

Rationale for irrigation

The provision of artificial water supplies guarantees the production of crops where water is the main deficiency. In the Nile Valley, the level of the floodwaters reflected the amount of water available for the farmers. Variations in level were reflected in variations in yield. The Nile Gauge noted by Pliny

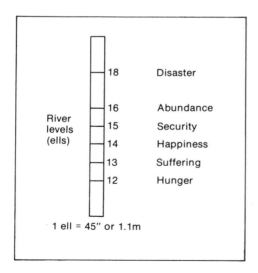

Fig. 13.3 The Nile Gauge or Nilometer. *Source*: Drower 1956: 539 after Pliny.
The gauge measured the height of the annual flood and indicated the impact upon the Egyptian society of the various levels

reflected the delicate balance between too little and too much water where societies were dependent upon irrigation for their basic food supplies (Fig. 13.3).

Crop yields (wheat and barley) under irrigation were a hundredfold greater than under rain-fed agriculture in the Babylonian civilizations. The control of the productive environment implicit in irrigation also allowed a wider variety of plants to be grown, including raw materials such as papyrus, reeds, and canes used in construction. In these controlled environments the first gardens were created. Ashur-nasir-pal II, Emperor of Assyria (884–859 BC) boasted:

> the meadow land by the Tigris I irrigated abundantly and planted gardens in its area; I planted all kinds of fruits and vines and the best of them I offered to [the God] Ashur my lord and to the temples of my land. (Drower 1956: 553–4)

Paradise, the Persian word for a walled garden, was located there and the Semitic tribes saw this environment as their Garden of Eden. For the Muslims the paradise was the irrigated environment of the oasis:

> On that day [of Judgement] there shall be radiant faces, of men well-pleased with their labours, in a lofty garden. . . . A gushing fountain shall be there, and raised soft couches with goblets placed before them . . . they shall feel neither the scorching heat nor the biting cold. Trees will spread their shade around them, and fruits will hang in clusters over them. (*The Koran*, Chs 88 and 76, Dawood 1956, pp. 30 and 17 respectively).

Not only was production virtually guaranteed and greater variety offered, but control of the total environment came within reach. The attractions of such resource management to peoples who had been at the mercy of the seasons and natural water shortages is obvious.

Hydraulic societies?

The surplus production afforded by irrigation affected the social organization of the groups practising this resource use. The effects of irrigation seem to have been in proportion to the scale of the water manipulation and the extent of communal activity necessary to establish and maintain the system. This relationship at the largest scale gave rise to what Wittfogel (1970) identified as 'hydraulic societies'.

The Wittfogel thesis

As a Marxist economist Wittfogel had been disappointed by the bureaucratization of government in Russia in the 1930s. In his studies of the history of irrigated agriculture he saw evidence of a link between the stratified irrigated societies and the totalitarian organization of the modern state. He divided irrigation systems according to their size, the small individually operated systems being 'hydroagriculture' while the 'large-scale and government-managed works of irrigation and flood control' created a hydraulic agriculture (Wittfogel 1970: 3). The latter, he suggested, grows up where sufficient water is available, where the sites require large-scale and unified rather than piecemeal development, and where, because of this size factor, a large investment of labour is needed. This labour has to be organized, jobs and responsibilities demarcated, and careful centralized and unified supervision provided.

The organization of the society to provide the labour for the initial construction and later maintenance of the irrigation works led, in Wittfogel's view to the creation of a totalitarian despotism, centred upon a king/emperor (often deified as a water god) supported by hierarchies of priests who interceded with the god for the people and who provided the bureaucrats to organize the labour on semi-military lines for the various tasks. The government of such a society existed by and for the irrigation system, illustrated by the district governors in the Egypt of the Pharaohs having the title 'Digger of Canals'. In this society importance was laid upon the ability to predict changes in the water supply such as the timing of seasonal floods, hence calendars and astronomical observations were developed; the allocation of water required measuring devices, the most usual being timing devices as the earliest types of clocks; canal construction encouraged the development of mathematics and geometry; and the increased potential value of the agricultural lands made land surveys essential.

The hydraulic landscape

The existence of such societies over several millennia led to the creation of what might be called a 'hydraulic landscape'. Use of land in such areas was very intensive. Not only were yields much higher than in the rain-fed agriculture areas, but multiple cropping throughout the year was possible and, where the irrigation waters brought silt, could be carried on without apparent loss of soil fertility.

In this 'controlled' environment a great variety of plants could be raised, including some brought from distant and different ecosystems. An earlier Assyrian emperor, Tiglathpileser I (1115–1102 BC), documented these transfers made possible by the irrigated agriculture:

> I brought cedars, box-wood, and oak(?) trees from the countries which
> I have conquered, trees the like of which none of the kings my
> forefathers had ever planted, and I planted them in the gardens of my
> land. I took rare garden fruits, not found in my own land, and caused
> them to flourish in the gardens of Assyria.   (Drower 1956: 553)

The first deliberate transfers of plants were under way, made possible by irrigation.

Within the hydraulic landscape were found the first cities, the clusters of dwellings of the rural labour force to which were added the barracks of the soldiers, the temples of the priests, the palace of the emperor, and the offices of the bureaucrats managing the irrigation system. These clusters of population, reaching over 100 000 inhabitants, were peaks in a densely settled countryside crossed by roads and navigable canals. It was a landscape where the hand of man was everywhere to be seen – from the straightened watercourses, the geometric fields, the tightly packed huts of the villagers, to the walls, temples, and storehouses of the cities. It was a landscape in which mankind had begun to buffer themselves against the vicissitudes of nature.

Yet the system was vulnerable and the Nile Gauge points the message. The origins of the biblical Flood were probably folk experiences from Mesopotamia translated into legend in the same way that the first heroic legend *The Epic of Gilgamesh* (3000 BC) dramatized human conflict with drought in the form of the 'Bull of Heaven' (Sandars 1960). The efficiency of the buffers against such threats depended in part on the efficiency of the central authority. As long as it kept the flows of water moving, the banks repaired, the canals cleaned out, and water allocations to schedule, the system worked. But any reduction in efficiency of a part tended to affect the whole. Warfare disrupted seasonal planning. For example the Mongol invasion in the thirteenth century AD indirectly caused the failure of irrigation in parts of Mesopotamia because it destroyed the regulating bureaucracy.

However, the system carried within itself the seeds of its own decay. Wittfogel claimed that, once established, the system was self-perpetuating with no time for innovations. With abundant labour there was no need to look to mechanization. Industrialization therefore was never developed. The irrigation system, using basically the diversion process, captured both water and silt. That silt accumulated on the fields (almost 1 m over 500 years in the Diyala Valley of Iraq) and over time the irrigation waters were less able to reach these raised areas. In addition, drainage problems arose with excessive flooding of the basins and salting of soils occurred. Productivity in these circumstances decreased, maintenance became difficult, canals were choked with silt, and no water reached the salted fields (Adams 1965). Even before the disruptions of the Mongol invasion there was evidence of problems developing in the irrigated areas. From 3500 BC to 1700 BC salt-tolerant barley had replaced the salt-sensitive wheat as the major crop in southern Iraq and yields of crops had declined over the period from 2400 BC to 1750 BC by almost two-thirds. It is claimed that such stresses led to the collapse of the Sumerian civilization (Jacobsen and Adams 1958). From 17 millions at the time of Nebuchadrezzar (605–562 BC) the population supported by the Tigris and Euphrates had sunk to 3 millions by the 1960s.

Irrigation since 1800

The situation in 1800

By 1800 irrigation systems had been developed on all the major continents except Australia and Antarctica. The irrigated areas were, by comparison with the contemporary figures, relatively small, perhaps at most half of the 44 million ha irrigated in 1900.

Concerning the details of location, in the Americas the diversion systems of the Indians of the southwestern USA had been adopted on the Spanish missions and haciendas of California and Mexico. In Mexico the 'floating gardens' still functioned but were much reduced in extent. In Peru the coastal valley diversion systems still existed but only as a shadow of their former extent, having been disrupted by earthquakes and siltation problems. In Africa the Nile basin system was paralleled by small deviation systems on the bajadas of the Atlas Mountains and Ethiopia, and the diversions from the Niger River on its inland delta near Timbuktu. Of indigenous irrigation systems prior to the Dutch arrival in southern Africa there seems little trace. In southwest Asia the Mesopotamian and Indus Valley systems were pot-marked by settled plains and abandoned canals as were the central Asian oases such as Ferghana, Bukhara, and Samarkand (Whyte 1961; Kovda 1961). In northern China the extensive floodplains of the Hwang Ho still received their seasonal watering. Beyond these arid lands, irrigation watered the tropical terraces and paddy fields of southeast Asia, the water meadows of Britain, European alpine pastures, and the polders of the northern European Plain. While it may be true that:

> in most part, each country has developed its own system of irrigation, in isolation, with only minor help or assistance from the experience gained in other countries. (Gulhati 1955: x)

the results in terms of technology were remarkably similar. By 1800, with the exception of overhead spray and drip systems, all the irrigation techniques of the 1980s were already in existence somewhere in the world.

Expansion, 1800 to the 1980s

Since 1800 the developments in irrigation have resulted mainly in the application of traditional techniques to wider areas, made possible by the advances in engineering techniques mentioned earlier. The expansion of the global area shown on Table 13.2 bears out Gulhati's comment (1955: viii):

> Most of the irrigated acreage has been developed during the last 80 years or so; during the next 20 years big developments are being planned all over the world and the irrigated acreage may increase to well over 500 million acres [200 million ha].

Locally there have been spectacular increases. In peninsular India the area irrigated in 1850 was estimated at 0.8–1.2 million ha. By 1905 it had reached 13.2 million ha, by 1945 19.6 million ha, and by 1955 (including Pakistan) 32 million ha, a twenty-seven-fold increase in 100 years – the equivalent of roughly 300 000 ha added each year. Most of these extensions in India and much of the global increases over the period resulted from the construction of large-scale reticulation systems drawing their water from large dams or barrages on permanent rivers. Since the 1950s this has continued to be the main technique for extending the irrigated area, although locally there have been important exceptions.

In the countries with a long history of irrigation, while new techniques have been applied to some of the old irrigated areas and, not least, to reclaim some of the old salted and silted fields, most of the techniques have been applied to lands irrigated for the first time. The result can be a complex pattern of traditional and modern techniques. In the Sudan the situation in 1955 is shown in Table 13.4. The expansion of area by 1965 was achieved by extension of the Gezira reticulation scheme and some pump irrigation. By 1978 further expansion had raised the irrigated area to 1.5 million ha. With 800 000 ha under irrigation the Gezira Board was:

> one of the largest agricultural enterprises under one management
> anywhere in the world. (Rainey 1977: 443)

Within the scheme population densities ranged from 40 to 150 persons km^{-2} in contrast to the surrounding sparsely populated grazing lands.

For parts of the global arid lands the main source of irrigation expansion has been the discovery of deep-lying aquifers which could only be tapped by the combination of improved deep-drilling techniques and the artesian

Table 13.4 Irrigation systems in the Sudan

	Area			
	1955[a]		1965[b]	1978
			10³ ha	10³ ha
	10³ ha	(%)		
1. *Diversion*				
Seil (Gash and Baraka rivers)	80.9	11.6		
Basins (along Nile)	41.7	6.0		
Sakiya	21.0	3.0		
2. *Reticulation*				
Pump schemes	146.9	21.1		
Gezira Scheme (from Sennar Dam built 1925)	404.7	58.3		(~800)[c]
Totals	695.3	100%	2509	1550[d]

Sources: [a] Gulhati 1955; [b] Highsmith 1965; [c] Rainey 1977; [d] FAO *Year Book*.

nature of the aquifer itself. Given this combination, overhead spray techniques by centre-pivot mobile booms (up to 560 m in length) have enabled field units of at least 98 ha to be watered from one well. On the central and southern Great Plains of the USA the Ogallalla sandstone aquifer supports a continuously expanding pattern of circular fields, visible from space on satellite imagery, which provide both irrigated livestock feed (maize, sorghum), industrial raw materials such as cotton, and human foodstuffs such as vegetables and sugar-beets. In new oases in the Sahara, such as Kufra (Libya), these fields provide both fodder for livestock and food for a farming community. The costs of such projects are escalating, however, as are the costs of more traditional schemes. With the increase in costs, economic, social, and environmental, has come increased government involvement.

The role of governments

The expansion of irrigation since 1800 has seen an increasing role of governments in the process of irrigation. Whether or not such a renewed role is a step back towards Wittfogel's hydraulic societies is debatable, but there is no doubt of the increasing central government commitments to irrigation in the nineteenth and twentieth centuries. In part this has been forced upon governments by the increasing costs of irrigation schemes. Over time the easiest sites for irrigation have been developed by private means, leaving the more difficult sites which usually require larger capital inputs. In part, however, the government interference also has been a deliberate policy, seeing irrigation as a means of developing national resources, especially in the arid lands.

Governments and the 'big dams'
In general the size and costs of the new irrigation developments in the arid lands since 1800 and particularly over the last 50 years have been such that private capital alone has been inadequate. With very few exceptions, the 'big dams' and associated reticulation schemes have only been possible with government funds and in some cases only with international funding. Thus, of India's 23.6 million ha irrigated in 1955, 14.4 millions were controlled (and had been built by) the government. Of these 14.4 million ha, 12 millions were irrigated by canals from dam storages, 1.8 million, from storage tanks and only 1 million from wells or *qanats* (Gulhati 1955). Conversely 4.8 millions of the privately controlled 9.2 million ha were irrigated from wells. Wells, it would seem, can be managed by individual farmers. Big dams need bureaucrats.

Even some of the modern wells, particularly those associated with centre-pivot schemes, may be so expensive as to require government financial support. The Kufrah Scheme in Libya, mining the aquifer in the Nubian Sandstone from depths up to 500 m to irrigate 15 000 ha, had cost LD 30 million

($US 8.8 million) from its inception as a government venture in 1973 to 1976, with annual maintenance costs of LD 10 million ($US3 million). As a pilot scheme to illustrate the problems of such developments it had value. As an economic project it was a failure (Allan 1976).

The Egyptian High Dam

The Egyptian High Dam Project has become the classic example of a massive government development with both positive and negative effects, some of which were not anticipated. Begun in 1956 with funds from the USA and the World Bank, the project – to raise the existing Aswan Dam to allow complete control of the annual Nile flood – was affected by the Suez War (1956) which led to withdrawal of Western financial support for political reasons. The USSR stepped into the gap in 1958 and over a further 10 years the construction was completed. In 1967 the filling of the storage began with the first complete impounding of the annual flood.

The forecast control of floodwaters was achieved, as was the extension of area irrigated by 0.7 million ha and the transformation of the 0.28 million ha of basin systems to perennial reticulation systems. Electricity from the dam doubled national energy production. The agricultural area was extended by 25 per cent with an accordant increase in production and the dam provided improved navigation and fishing resources. The Abu Simnel Temple was rescued from the rising dam waters by United Nations funds, and displaced villages were resited on newly irrigated lands.

On the negative side, the control of flooding stopped silt-borne nutrients reaching either the fields or the fishing ecosystems of the delta and adjacent Mediterranean. The results were the collapse of the fishing industry and high annual costs for artificial fertilizers for the fields. Further, population growth outstripped the additional food supplies from irrigated lands as growth rates were virtually 3 per cent per year over the period 1960–70, when agricultural production increases averaged only 2.5 per cent per year (Holz 1968; Shibl 1971). By the late 1970s the food needs had to be met by massive imports of wheat from sources such as Australia. A medical problem also had been intensified as the extension of permanent canals led to a population explosion in the *Bilharzia*-carrying snail, leading to outbreaks of the debilitating disease among the rural population. Irrigation had been extended but at a price.

Case study: Irrigation in the USA (1500–1980)

Developments 1500–1900

Prior to the Spanish occupation in southwestern USA, Indian groups had established various diversion systems. These included planting crops on lands which were flooded naturally – such as the Mandan and Pawnee

Indians' use of alluvial flats or bottom lands on the eastern edges of the Great Plains and the Hopi planting of several run-on areas in hopes that at least some would produce crops. Floodwaters were diverted out of the arroyos (wadis) on to alluvial fans by the Zuni Pueblo peoples; while elaborate canals – such as the 30 km system of the Hohokam peoples had been developed AD 1200–1400 in the Gila River Valley (Armillas 1961). The Spanish found these latter systems in ruins, but whether from salinity or silt problems, or warfare, is difficult to say. The Spanish added their own terracing and canal techniques learned from contact with the Moors in Spain. Small diversion schemes seem to have served most of the missions and military posts which were still being established in California at the time of the American War of Independence in the east.

From the Anglo-American occupation of the southwest from about 1850 to approximately 1900 various small-scale irrigation schemes were introduced by individuals or private groups (Logan 1961b). Several group settlements chose irrigation as the means to provide subsistence foodstuffs. The Mormons, a religious group seeking a refuge in the newly acquired territories, settled on the foothills of the Wasatch Range overlooking the Great Salt Lake in 1847 and used the bajadas to site their villages, each with its diversion irrigation system (Nelson 1952). Other social groups, such as the German Colonization Society and the Greeley Union of the 1860s, which settled at the foot of the Rocky Mountains on the western edge of the Great Plains, saw in irrigation the means not only by which arid lands could be made productive, but also a system of resource use which required co-operation and not only philosophical but also practical cohesion. Where the groups had strong leadership, practical abilities, and a market for produce, they survived, but many fell victim to internal disputes, bad management, lack of irrigation expertise, and isolation from markets (Brown 1948).

In California, group schemes such as the German settlement at Anaheim prospered, and several private companies were established to provide water for farmers on lands subdivided out of the ranches which previously covered the better-watered valleys. Having achieved local successes the Chaffey brothers were invited to Australia in 1885 by the Victorian Colonial Government to apply their methods to the Murray River Valley. Their efforts, although marred by disagreements over the terms of their contract and technical problems of getting the river water up out of the valley into the fields, set the foundations for irrigation in Australia (Rutherford 1964).

By 1900, 2.8 million ha were under irrigation in the USA, the bulk lying in the western arid states and over half of that area having been irrigated only within the previous decade (Table 13.5). By 1900, however, there were signs that the private sector monopoly of irrigation development was soon to be challenged. The most convenient sites were being rapidly occupied and the size and cost of projects were escalating.

In 1902 one of the last major private irrigation projects was begun – the

Table 13.5 Irrigation in the USA

A. Area under irrigation

Date	Area (10^6 ha)
1889	1.4
1900	2.8
1909	4.7
1929	6.8
1939	7.4
1949	9.1
1958	12.2
1968	15.7
1978	16.7

B. Regional irrigation

Date	Percentage of national irrigated area	
	28 eastern states	*17 western states*
1955	8	92
1964	38	62
1974	10	90[a]

[a] 17 western states and Louisiana.

Sources: Ruttan 1965; FAO *Year Books*; US Bur. Census, *Stat. Abstract of the US 1979*, Govt. Printer, Washington.

Alamo Canal, diverting water from the Colorado River below Yuma, Arizona, by a loop through Mexico then into the Salton basin of the Imperial Valley of California, a distance of over 90 km. By 1904:

> nearly 8000 people had settled in the valley, 700 miles [1120 km] of canal were in use, and 75 000 acres [30 000 ha] of land were being farmed. (Dunbier, 1968: 177)

The project proved costly to operate since the inlet from the Colorado River needed constant dredging to remove the accumulating silt. Levees for flood protection also were needed. In 1905 those levees broke under a major flood and the whole Colorado River passed down the man-made channel into the Salton basin for 16 months, filling the arid basin to a maximum of 21.6 m and creating a lake of 132 000 ha before the gap in the levees could be plugged. Each year thereafter, however, until the government projects of the 1930s affected the river regime, another 12 300 cm³ of silt was deposited along the lower Colorado and each year the flood levees had to be raised. The schemes were obviously becoming too complex and too costly for private development alone.

205

Developments 1900–1960s: the federal phase

By 1900 the remaining federal lands were mainly in the arid areas and were not attracting settlers even as free land (with residence and improvement requirements). One means of ensuring disposal of such lands would be to provide a water supply. If the government could do this, it would at once dispose of the land, encourage resource development, and establish an agricultural population on the land. The provision of water by the federal government would also ensure a more equitable distribution among prospective users, without the risks associated with private monopolies of the water supplies. Such were the arguments for government intervention at the turn of the century. Subsequently, concerns to control flood hazards, to encourage conservation of resources, to provide cheap electricity for industry and recreation areas for the general public have been added to the list.

Government intervention in irrigation was made legally possible by the establishment of the 'Reclamation Service' as part of the US Geological Survey in 1902. By 1907 it had become a separate body. In 1923 it was renamed the Bureau of Reclamation and in 1980 the Water and Power Resources Service. The latter:

> reflects a broadening of the initial purpose of reclaiming arid western lands to responsibility for water and power resource management.
>
> (AAG 1980: 27)

The results were a series of spectacular dams built along the major rivers crossing the arid lands of the western states. The most modern engineering techniques were applied and the price of the dams and canals written off as public expenditure. The only charge on the water supplied was the cost of maintenance of the system. Most of the dams were multi-purpose, providing irrigation and domestic water, hydroelectric power, flood control, some navigation, and most recreation facilities. Within 30 years the area under irrigation had trebled and by the late 1970s it had increased sixfold (Table 13.5A), while the dominance of the western states was re-established (Table 13.5B).

Developments post-1960s: competition and complications

Up to the 1960s most of this increase in irrigated area was the direct result of government projects. From the 1960s onwards private projects using centre-pivot systems based on underground supplies have added substantial areas. Most of the latter, however, are 'mining' operations. For example, the draw-down on the Llano Estacado of Texas from 1949 to 1968 took the water table from 12 m to 30 m deep and the supplies are estimated to run out by the year 2000.

There is no doubt that without these massive investments of national funds the western states would have remained economic backwaters with sparse

Table 13.6 Problems of managing the Colorado River water resources in the southwest USA

Problem type	Explanation
1. *Natural events*	
1.1 Variability of rainfall in river catchment	Planning for storage facilities in the 1920s apparently coincided with wetter phase of precipitation (rain and snowfall); now is drier phase and storages are inadequate
1.2 High river silt load	Storages on river trap silt and their capacity over time is reduced. Lake Mead's life span originally forecast for 50 years before silt filled it, has been extended to about 100 years because Glen Canyon Dam upstream now catches the silt!
1.3 Flood peaks associated with scouring and deposition in canyons	Control of floods has prevented renewal of deposits on sand bars in canyons used as campsites for recreation voyages, now they are not being renewed, only eroded
2. *Human activity*	
2.1 Removal of water has reduced river flow along its course and to zero at mouth	Source of inter-citizen/private company disputes over water rights. This in the 1970s included Indian claims for water rights on their reserves Source of inter-state disputes over water rights to the Colorado itself Source of international dispute between USA and Mexico over US guarantees to Mexico of minimum flow by Treaty of 1944.
2.2 Removal of water from, plus return of saline drainage water to, the river has affected quality progressively downstream	Led to loss of agricultural land by increased soil salinity Led to legal disputes at all three levels noted for 2.1 above on the quality of water provided. Currently the US Government is constructing a desalination plant at Yuma at cost of about $200 million, with annual operating costs of $10 million to guarantee Mexico its treaty rights (2.1 above) to the $1850 \times 10^6 \mathrm{m^3 yr^{-1}}$ quantity and at least 140 ppm quality.
2.3 Reduction of available quantities and quality of water has led to reassessment of uses	Reassessments have tried to evaluate the various benefits of alternative water uses

Sources: After Peterson and Crawford 1978 and field data 1979.

populations. The impact of the irrigation schemes, however, has been both positive and negative. The western states, benefiting from expanded water resources as a result of the government projects which have both provided new supplies and guaranteed supplies to the private schemes already established, experienced a population boom from the 1960s onwards, being some of the fastest-growing states in the nation. During the 1960s while New Mexico only grew by 7 per cent, Utah grew by 19 per cent, Colorado by 26 per cent, southern California by 27 per cent, and Arizona by 36 per cent (Peterson and Crawford 1978).

The growth of population meant that alternative demands for water increased and the capacity of the Colorado River system to provide water was being rapidly exhausted. In the late 1970s, the river no longer reached its delta in the Gulf of Baja California, all the water having been impounded and drawn off upstream. The results were a complex array of problems (Table 13.6). They range in scale from disputes between individuals over water rights to legal battles between states and negotiations between nations. As water has become an increasingly scarce resource, its major use for irrigation purposes in the western USA has come under attack.

The problem of alternative water use is twofold. First, irrigated agriculture requires enormous volumes of high-quality water as we saw in Chapter 5. Second, alternative uses not only require less water but provide greater economic return and job opportunities (Wollman 1962).

By comparison with other parts of the USA the use of irrigation water in the arid southwest is relatively inefficient in the economic sense. In the 1950s the contrast in use of irrigation water between the western and eastern USA and the associated lower return per unit area were obvious (Table 13.7). The paradoxical situation had arisen where irrigation was apparently paying better in the humid than the arid parts of the nation. In the west the extensive areas under hay and forage, associated with the ranching operations, contrasted with a larger area under high-value speciality crops in the eastern states. At this time an argument could have been put forward that the vast federal schemes were mainly providing cheap summer feed for the western ranches.

The expansion of population in the western states from the 1950s onwards was achieved by relatively small expansions of irrigated land and much greater consumption by the more efficient non-agricultural users. In Arizona as an example, the increased use of water from the 1940s to the late 1960s was associated with a slight fall in irrigated area and substantially greater increases of income from non-agricultural users (Table 13.8A). The marked contrast in volumes of water needed to return $1000 value of output also was obvious (Table 13.8B). For every agricultural worker supported by irrigation developments, sixty to seventy industrial workers could be supported by industrial use of that water.

In the light of such figures it is not surprising that the proposals of 1966 for the grandiose North American Water and Power Alliance Scheme

Table 13.7 Use of irrigation water in the USA (1954–60 averages)

Category	Percentage of irrigated area in each crop	
	28 eastern states	*17 western states*
A. *Crops*		
Food and feed grains	26	31
Hay and forage	16	40
Cotton	8	9
Speciality crops: fruit, vegetables, potatoes, tobacco, sugar-beet	44	15
	100	100
B. *Value of production per acre*	*$ per acre*	
	28 eastern states	17 western states
Minimum	169–254	11–17 (Colorado River schemes)
Maximum	727–1036	413–18 (southern California)

Source: Ruttan 1965.

(NAWAPA), by which the expanding water demands of the southwestern USA were to be met by massive transfers of 'surplus' water from Alaska and the Canadian Yukon Territory have been shelved. Opposition had come from conservationists concerned at the ecological impacts at both source and destination areas and Canadians opposed to export of a valuable resource. Further, the costs (estimated at $80 million over 30 years) of transferring such water when it was to be used mainly for relatively inefficient agricultural use were considered excessive, and opponents suggested as alternatives more efficient use of existing supplies and more effective pricing schemes to curb the future demands. The increasing scarcity of the resource brought pressure to rationalize use and prevent further unlimited irrigation expansion.

As the most highly developed technologically oriented nation, the evidence from the arid lands of the USA points to a restricted future for irrigation in the arid lands. What of the global view?

The future of irrigation in arid lands

There is no doubt that existing irrigated areas are making a vital contribution to the livelihood of the global arid nations. Whether as producers of vital

Irrigation: panacea or Pandora's box?

Table 13.8 Water use in Arizona, 1940–1968

A. The changing use of water 1940–68

Criteria	1940	Changes	
		1940 to early 1950s	early 1950s to 1968
Water used (million acre/feet per year)	2.95	+ 3.58	+ 1.27
Area irrigated (000 acres)	665	+ 635	− 96
State income ($million per year):			
Agriculture	66	+ 129	+ 5
Mining	45	+ 14	+ 66
Manufacturing	25	+ 37	+ 518
Government	106	+ 139	+ 899
Others	272	+ 350	+ 1600
Sub-total	514	+ 669	+ 3088
State population (000s)	499.3	+ 395.0	+ 845.0

B. Water needs by production sector *c.* 1968

Production sector	Direct water needs per $1000 output (acre-feet)
Forage crops	46.9
Food and feed grains	42.1
Citrus fruits	11.2
Fruit and tree nuts	10.0
Vegetables	6.4
Mining	0.15
Canning, preserving, freezing	0.04
Fabricated metals and machinery	0.003

Note: Values indexed to 1950s.
Source: Kelso *et al.* 1973.

foodstuffs, raw materials for industry, or export income, their role is essential to the welfare of the nations.

There also is no doubt that the technology exists to extend irrigation to new agricultural areas. On a global scale some of the greatest potentials probably lie in the central Asian steppes. If the Siberian rivers (Ob and Yenisei) are ever to be diverted south to supplement the current Amur and Syr Darya schemes, an area of perhaps 2.8 million ha could be added to the 12 million ha of the 1960s (Hollis 1978). There are also areas where diversion and reticulation schemes could add irrigated lands even within the traditional 'hearths' of Mesopotamia and the Indus Valley, *if water could be used more efficiently*. The problems are not technical, but rather whether the returns

210

justify the costs – economic, social, and environmental.

As an illustration, the proposals for diversion of the Siberian rivers south into the interior drainage basins of the Aral and Caspian Seas are seen not only as providing additional water for irrigation but also some relief for one by-product of existing irrigation schemes – the catastrophic falls in levels of these two water bodies. Since 1930 the level of the Caspian Sea has fallen 2.3 m, while between 1961 and 1973 the Aral Sea level fell 3 m! This fall is the result of reduced river inputs as water is diverted for new irrigation schemes. As a result, fish catches have declined, salinity levels have risen, and navigation has become more difficult. Part of any new water inputs may have to be as compensation flows to stop these falling sea levels (Hollis 1978).

Whether further irrigation schemes in the arid lands will be economically successful is debatable. The escalating costs of construction and maintenance of the schemes are not being offset by equivalent increases in the prices of irrigated produce. In addition, demand for water is increasing in the humid lands and less water will be available for the arid areas in the future. For many arid nations expansion of the irrigation areas may not prove a sound economic investment.

The social costs of irrigation are less easy to document, but reflect the change in population growth rates and locations from the nineteenth to the twentieth centuries. Seen by governments in the nineteenth century as a traditional and guaranteed method of establishing a pattern of close rural population in the arid lands, by the mid-twentieth century the economics of irrigation required large farms, heavy mechanization, and a much reduced seasonal harvest labour force. From the government viewpoint investment of funds in industrialization would appear to provide more jobs and more income for the citizens. Thus, any proposals for large-scale government investment in future irrigation schemes will have to weigh very carefully the 'opportunity costs' of alternative water uses not only in economic but also in these social terms.

The environmental costs of irrigation are now recognized to be considerable. Salinity problems seem to be inevitable unless efficient (and costly) drainage schemes and careful control of water applications are provided. The evidence of past mistakes can be found in salted barren waste lands, which once were productive fields, whether in traditional 'hearths' of irrigation such as Iraq and Pakistan, or the Colorado Valley of the USA where the affected schemes may be less than 30 years old (Table 4.4). Pessimists point out that many of the new irrigation areas of Soviet central Asia are less than 10 years old. The lack of any salinity problems so far is no guarantee for the future.

For the global arid lands the irrigated areas will continue to be a major source of income and national welfare, but it is unlikely that any major expansion of the area will be achieved without massive financial investment. Of necessity, this investment will probably come from more humid lands and, increasingly, planners will argue that the water might be used better even for irrigation in those humid lands to intensify existing humid land agriculture or for alternative non-agricultural uses. In addition, within the arid lands, the current mining of underground supplies will probably result in the local abandonment of irrigated lands as supplies dwindle or become more costly.

CHAPTER 14

Mining

Mining as exploitive resource use

So far in the discussion of resource management we have considered non-exploitive systems of resource use. However, several systems have resulted in exploitation and the run-down of the resource base through mismanagement and/or ignorance of the nature of the resource and an inability to minimize environmental impact. The exploitation of the resources, in the sense of their destruction or exhaustion (see Fig. 9.1), was not the intent of their use, whereas such exhaustion is the implicit effect of mining if not the stated aim of the miners.

Mining of mineral ores implies the processing of those ores for use usually at some location remote from the mine (see Fig. 10.3). In the arid lands it usually implies the removal of the ores some considerable distance, frequently out of the arid lands, to the market or for further processing. As an exploitive form of resource use, the exploitation may be relatively short-lived – depending on the size and rate of consumption of the ore body or the fluctuating prices for the ore or costs of extraction. In spatial terms the mines usually are point locations around a limited number of shafts or open-cast excavations and in the arid lands may represent a node of high capital investment, population density, and generated income amid a vast 'empty' area of little or no other significant resource use (Pl. 14.1). The local impact of such mining activities, therefore, apart from at the node itself, may be relatively insignificant. The main impact in economic terms may be at regional or national level where income from royalties is fed into the national treasury and service industries provide support equipment, transport, and supplies.

The mining resource-use system requires a mine site, supplies of labour, power, basic services, and a communication/transport system to convey the product to market. Depending upon the ore, the processing may involve: the actual mining of the rock, the concentration of the rock into mineral ores by removal of useless rock (e.g. pelletizing of iron ores), and finally, the smelting and refining of the ore for immediate use. Since this latter process

213

Plate 14.1 Peko Mine, Tennant Creek, Northern Territory, Australia, photographed 1971.
A copper-mine settlement with ore stockpiles on the left centre, tailings-pond top centre, mineshafts and associated buildings at centre, single men's huts lower centre with canteen at right, and family houses top right. A 'boom town' site in the mid-1960s, at the time of writing in 1981 this mine had been closed down as a result of the declining price of copper.

requires large amounts of energy the location of this stage may be closer to the market or source of energy.

Mining the raw materials

Despite the harshness of the surface environment the minerals of the arid lands have been exploited successfully for several thousands of years. At either end of the spectrum of bulk:value ratios (from salt to copper and gold) there is evidence of mining from earliest historical records. Trade routes for salt and slaves crossed the western Sahara, and the Sinai Peninsula was producing copper for Egypt as early as 2600 BC (Bromehead 1956). Even in the 1970s these are still important products from the arid lands, for while the core and predominantly arid nations provide only 7 per cent of the world's copper, they provide a third of the world's salt and three-quarters of the gold. When the semi-arid and peripherally arid nations are included, virtually 80 per cent of the world's copper and salt and 89 per cent of the gold production originates in the arid nations (Table 14.1). Considering the role of salt as a raw material in the chemical industries and copper in the electrical goods industries, these are vitally important industrial raw materials, while gold has remained a hedge against global inflation.

214

Table 14.1 Mineral production in arid nations

Minerals

Arid nations	Gold 1973–7 ave. (000 kg)	Diamonds 1977 (industrial and gem.) 000 metric carats	Copper 1977 (000 tonnes)	Lead 1977 (000 tonnes)	Bauxite 1977 (000 tonnes)	Salt 1973–7 ave. (000 tonnes)	Iron ore 1973–7 ave. (000 tonnes)
Group I 100% arid	0	0	4.8 (Mauritania only)	0	4600 (Saudi Arabia only)	5 820.6	6 692
Group II 75–99% arid	17 135.6 (Australia only)	4 661	289.9	617.9	22 506 (Australia only)	49 876.8	59 759
Group III 50–74% arid	748 579	8 033 (S. Africa only)	239.0	37.8	567 (Turkey only)	2 171.6	11 322
Groups I–III (% of world)	765 714.6 (74.5)	12 694 (31.5)	533.7 (6.7)	655.7 (20.1)	27 673 (34.8)	57 869 (35.4)	77 773 (15.5)
Group IV 25–49% arid	45 735.7	746	2644.7	833.9	5 168	50 596	116 513
Group V 1–24% arid	99 170	13 558	3119.7	1020.1	5 856	22 268	235 479
Total arid nations (% of world)	910 630.3 (88.6)	26 998 (66.9)	6298.1 (78.7)	2509.1 (76.8)	38 697 (48.6)	130 733 (79.9)	429 765 (85.6)
World	1 027 400	40 366	8000	3270	79 600	163 530	502 060

Sources: Statesman's Year Books and UN Year Books

Fertilizers from the 'desert'

Apart from common salt (sodium chloride) a variety of other chlorides, sulphides, oxides, and phosphates are being mined in the arid lands as fertilizers for agricultural application – often in more humid areas. Phosphate rock (calcium phosphate) as a source of superphosphate is one such basic raw material and 70 per cent of the world's known reserves are found in arid northwest Africa (Shreiber and Matlock 1978).

Most of the playas and salt lakes are potential sources of mineral salts. The Dead Sea waters have been evaporated for their salts since 1932. Potash plants are currently supplying 'life' in the form of fertilizers for both Israeli and Jordanian agriculture and various salts for the chemical industries (Karmon 1971).

One resource which the coastal arid lands offered at the mid-nineteenth century was the guano deposits – the dung of the millions of birds nesting on the islands. Off the coast of South West Africa, discovery of 22.5 m thick deposits on Ichaboe Island in 1828 had to await the discovery of the role of chemical fertilizers in plant growth by Liebig in the 1840s. Within months of the recognition of the role of phosphates in agriculture a rush to mine the deposits began, and by the 1850s had stripped the island of most of its 800 000 tonnes of guano (Brittan 1979).

Much earlier the Inca civilizations had used guano from islands off the coast to fertilize irrigated fields in the valleys of the Peruvian Desert. As with South West Africa the boom in exploitation came in the mid-nineteenth century, but unlike the African situation the Peruvian production has continued – in theory under government control and at a rate approximately equal to the rate of accumulation. With each bird (cormorants, boobies, and pelicans) producing about 45 g of excrement per day, the average bird population of 5 million would produce about 82 million kg of guano per year, a proportion of which would accumulate on the islands (Paulik 1971). This would be, along with salt evaporation from sea-water, one of the few examples of mining a renewing resource.

Base metals

For the major base metals (iron ore, bauxite, and lead) the arid nations contribute a significant proportion of global production (Table 14.1). For iron ore the core and predominantly arid nations contribute 16 per cent (of which Australia provides 11%), while the semi-arid and peripherally arid add a further 70 per cent of the global figure (of which the USSR provides 25%). Three nations, Saudi Arabia, Australia, and Turkey, provide just over a third of the world's bauxite, with the other arid nations only adding a further 14 per cent. Australia alone contributes 28 per cent of the total, and also dominates the lead production from the predominantly arid nations, with 13 per cent of their 20 per cent share of the global production. The arid nations as a whole provide just over three-quarters of the global lead production.

Precious metals and gems

Production of gold from the predominantly arid nations provides over three-quarters of the world output, the bulk coming from South Africa (73%), while the remaining arid nations provide another 14 per cent. Opal production is virtually limited to the arid areas of Australia which provides some 95 per cent of world production. Diamond production is dominated by South Africa (20%) and the USSR (25%), with the arid nations contributing two-thirds of the world output. New discoveries in semi-arid northern Australia promise to reinforce this dominance.

Problems of mining in the arid lands

To the 'normal' problems facing the miner anywhere in the world the arid lands add several fundamental difficulties which may be grouped under two headings. First are those resulting from the remote location, lack of knowledge, and accessibility of the arid lands. Second are those related to the high processing costs in the arid lands.

Remoteness and accessibility problems

Exploration
Until the development of aerial geomagnetic surveys and off-road vehicles, knowledge of the geology of the arid lands was limited. Only in the last 20–30 years has detailed information of the core arid areas of North Africa, Arabia, and Australia begun to be accumulated. With increased knowledge the chances of discovery of deposits have increased. Indeed, the Australian economic historian, Geoffrey Blainey (1969), suggested that the mining discoveries of interior Australia were a function of the increased knowledge of the terrain acquired incidentally by the pastoralists who replaced the Aborigines in the latter half of the nineteenth century. The search for one mineral may produce others. The Mt Isa copper, lead, and zinc deposit in Queensland, Australia, was found by gold prospectors, and oil searchers in the Sahara discovered the Nubian Sandstone aquifer in the 1960s on which much current irrigation is based.

Accessibility
Discovery of remote mineral formations, however, does not guarantee a mine. The costs of transport over long distances to market means that the price of the ore must be high, and the quality of the ore higher than fields closer to markets. Since most of the markets are in the humid lands this means that ore bodies in the arid lands must be of high quality to justify the abnormally high costs of access and transport. Only copper and gold could lure the Egyptian miners into the Sinai Peninsula (2600–1200 BC) and the Nubian Desert (1292–1225 BC) (Bromehead 1956). In the 1850s copper of

217

27 per cent purity barely paid the cost of horseback transport from the Blinman Mine in the Flinders Ranges of South Australia to the coast 219 km away for export to Britain.

All the iron-ore mines opened up in the 1960s in northwest arid Australia had to be provided with rail links to the coast where the ore was concentrated before being shipped to the market – mainly Japan. All the major phosphate mines in West and northwest Africa are linked by rail to the coast for the same reason. The Australian links were parallel tracks up to 400 km long, the African links ranged from 25 km to 400 km. In one case, Bu-Craa in the Western Sahara, a mobile sand-dune complex blocked access to the coast and the ore goes out by a 100 km conveyor belt (Shreiber and Matlock 1978). Apart from the engineering problems of such links, the cost of the construction has to be borne by the developer. Unlike deposits discovered in the humid lands, there is not likely to be any pre-existing settlement structure or transport network into which the mine production can be channelled.

Labour

The remote mines in the arid lands offer the worker all the hazards of normal mining operations in a harsh environment of high temperatures and high dust levels. To provide labour, developers have to resort to either the stick or the carrot. 'King Solomon's Mines' for copper (*c.* 1000? BC) in the Negev were worked by a slave labour force, as were the Egyptian copper mines in the Sinai 1000 years earlier. In the twentieth century the inducement is usually monetary, with base wages being supplemented by remote area allowances or bonus payments. Even so, the turnover of the workforce tends to be higher here than in mines less remote from the settled areas. In the iron-ore mines of northwestern Australia annual turnover rates range from 300 per cent to 500 per cent (i.e. each job is filled by three to five different workers during an average year).

Infrastructure and maintenance

The remote locations require that developers bear the costs of basic support services and face high costs for importing goods and services for the workforce. The result is often the creation of a 'company' or mining town solely to serve the labour force of the mine, a cost not usually required in more humid lands. For the iron-ore developments of northwestern Australia in the 1960s, the capital costs were approximately $A800 million, of which $A500 million went to provide the infrastructure (railways, port facilities, new towns and roads).

Provision of water for mining operations and for the mining population is an obvious problem. The development of the goldfields of Kalgoorlie in arid Western Australia in the 1890s was only possible after the state government built the 562 km water pipeline from a catchment area in the humid coastal highlands. A two-year drought (1944–6) exhausted the reservoirs of the silver, lead, and zinc mining town of Broken Hill, Australia. Water had to be imported by railway tanker wagons from the River Darling 100 km distant. The town's 27 000 inhabitants needed up to eight trains per day each carrying 409 m^3 of water over the period from August 1944 to 17 January 1946. On that final day the last train – typically for the arid lands – was derailed by floods! The cost of the railed water (paid for by the mining companies) was four times the normal local price, which already was several times higher than the metropolitan capital Sydney's water price.

Other basic processing problems face all arid lands mines. Not only water for the labour force but also for the preliminary crushing of the rock may be needed. In the Kalgoorlie goldfields in the 1890s the standard and traditional 'washing' of the alluvial rock to separate out the gold practised in the more humid areas could not be used. 'Dry-blowing', by which the crushed rock was winnowed by air currents, was adapted. Timber for the mine shafts, buildings and fuel for steam-engines, smelters, and domestic use took their toll of local vegetation and in many cases had to be imported long distances (Blainey 1969). In all cases the costs were higher in the arid lands. Again the moral was that returns would have to be proportionally greater to carry this extra cost.

Case studies

'King Solomon's Mines'?

Archaeological evidence of copper mining and smelting in the Negev Desert was first noted in 1934 in an article in the *Illustrated London News* of 7 July on 'King Solomon's copper mines'. Evidence was provided of adit mining at three sites where the rock was first roasted to concentrate the metal and then transported in panniers on mule-back to the coast near the present Eilat. Here a smelter built in the centre of the Wadi Arabah (the biblical 'Valley of the Smiths') used the prevailing northerly winds to provide a forced draught for the charcoal-burning furnaces. The labourers were slaves as indicated by the fortified barracks and military camps associated with each site. Once refined the metal must have been transported by sea to Suez then overland and again by sea to its final destination (Glueck 1959). Subsequently, the age of this mining complex at Timna has been put back to at least 4000 BC, with Egyptian control from 1400 to 1250 BC, but no direct evidence of operations during Solomon's reign.

In 1959 the Israeli government reopened the Timna site. The contemporary copper prices and military necessity made the 1–2 per cent pure ores attractive. As done 3000 years previously, preliminary concentration of the rock at the minehead to a 'cement' of 80 per cent purity preceded its export by motor truck to Eilat (Karmon 1971). A volunteer but highly paid labour force of 600, rising to 1000 in the mid-1970s, partly resident in Eilat, had replaced the earlier slaves. Production of 900 000 tonnes of rock gave 10 000 tonnes of copper per year after refining, and an export trade to Belgium, Spain, and Japan was built up. In 1975, however, a fall in world copper prices to less than £900 per tonne brought closure of the mine and 'King Solomon's' copper was once again abandoned (Karmon, 1976).

Diamonds in the Namib

Geological erosion of some 800–1500 m of the surface of what is now the High Veld of northern South Africa had left both diamond-bearing rocks exposed near Kimberley and produced a train of alluvial deposits along the main drainage system – the Orange River – to the sea. There the northerly flowing Benguela Current had distributed the gems along the coastline in marine terraces north of the estuary.

The discovery of diamonds at Kimberley in the 1860s brought a 'rush' of European miners to work the surface alluvial deposits. As costs of mining the deeper ore veins rose, companies were formed. By the 1890s one company, De Beers, had a controlling interest, supplying 90 per cent of world demand. Discovery of the alluvial deposits in the Namib Desert did not occur until 1908 when a railway worker on the newly constructed German colonial railway inland from Luderlitz found valuable specimens and another diamond-mining rush developed. Since the discoveries were on government land the colonial development company, which had been formed to build the railway, was given sole development rights over a strip of desert 160 km wide north from the mouth of the Orange River to latitude 26 °S. Using cheap African labour to hand-sift the surface sands, which had been deposited on top of the marine terraces, and with all production controlled through a new diamond market in Berlin, profits were rapid and enormous. At the peak of production (1913–14) company dividends were 3800 per cent!

The occupation of German South West Africa by South African troops during 1915 and the confiscation and then purchase of German assets by the Anglo-American Corporation of South Africa (under Ernest Oppenheimer) consolidated this rival producer. With further diamond discoveries at the mouth of the Orange River in 1928, Oppenheimer was able to obtain control of De Beers and begin the monopoly of the diamond production which, although no longer complete, is still a strong market force to maintain prices in the 1980s.

After closure during the World Depression in the 1930s, the fields reopened in 1935 and massive earth-moving machinery is now removing the coastal sand dunes to get at the marine terraces beneath. The ratio of volume of useless overburden removed to productive gemstones is some 2000 million to one, the highest such mining ratio in the world, and is an indication of the value of the stones themselves. The diamond fields operate some of the largest earth-moving equipment in the world and in 1977 produced almost 2 million carats, which although only 5 per cent of the global production of both industrial and gemstones, was 20 per cent of the global gem production (Brittan 1979). From the 'deserts' comes not merely wealth but the jewellery to adorn it.

The significance of mining

Despite the high costs of development, mining has been undertaken and is being undertaken more extensively in the arid lands. The significance can be assessed in economic and environmental terms at several scales from the local to national and international levels.

Economic significance

For many sites in the arid lands the mines represent the first and only relatively permanent centres of resource use, particularly in the desert mountains. They offer jobs to pastoral nomads (some 15 000 bedouin being employed by ARAMCO in Saudi Arabia alone), local goods and services otherwise unavailable (retailing, health and welfare services) and new transport routes. In some cases the water supplies brought in for the mines allow limited irrigation development for the local market. At particular sites, however, this local impact can be relatively small for the mining centres might only have populations of 1000–2000 and the population may have to depend upon distant larger centres for some health, entertainment, and welfare services.

At the regional or national level, the income from mining royalties and the provision of services may make a substantial contribution to income which may be recycled to improve general living standards. The nineteenth century gold-rushes extended from the humid into the world's arid lands in the Americas, South Africa, and Australia. Some of the profits from these operations became the capital for subsequent private and official land settlement projects in both humid and semi-arid lands. Within the core arid nations, however, the most spectacular effects have come since the expansion of oil production from the 1950s onwards.

Case study: Bahrain

Prior to the discovery of oil in 1932 the low limestone and sandstone island in the Persian Gulf provided a precarious living for its 100 000 inhabitants from small areas of date palms, fruit and vegetable gardens supplemented by fishing, pearling, and trading (Fig. 14.1). The 62 mm annual rainfall barely filled the domestic water cisterns and gardens and palms had to be irrigated from shallow wells. Over the rest of the island the burial mounds

221

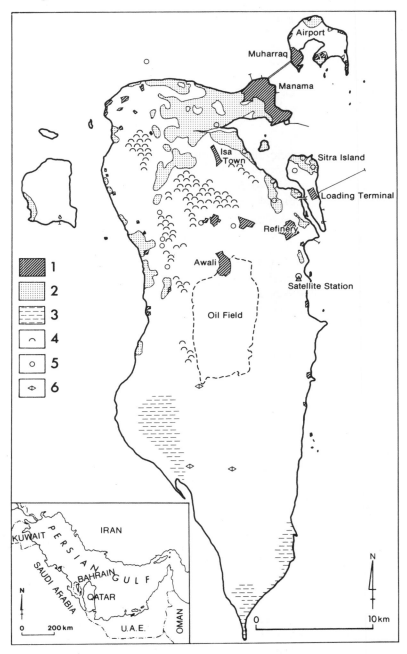

Fig. 14.1 Bahrain. *Source*: Guide Map of the Bahrain Island in Belgrave 1975
Key: 1 – Urban areas; 2 – Gardens; 3 – Salt flats; 4 – Burial mounds; 5 – Springs; 6 – Sheikh's hunting lodge

of at least 5000 years of continuous settlement went relatively undisturbed.

After 5 years attempting to collect sufficient capital to prospect, a New Zealander, in association with the Standard Oil Company, discovered oil in 1932. This led to the first exports and the construction of a small refinery in 1935. Artesian water was discovered by the oil-drillers and agriculture was intensified, but it has been oil production, rising to 2.5 million tonnes in 1979, which has provided the economic base for the kingdom which became independent in 1971 after 110 years as a British protectorate. By 1974 oil revenues provided 78 per cent of the national income for the 250 000 population and had been used to establish basic services such as roads, hospitals, free medical care and education, electricity to all households, a new settlement of 15 000 population (begun in 1968) at Isa Town, and an international airport (completed 1971) handling about 86 000 passengers per month in the mid-1970s. The population grew from 182 000 in 1965 to 341 000 in 1978 (comprising 88 nationalities) mainly through immigrant labourers from the surrounding Gulf States and Pakistan. The oil revenues have both improved the national economy and standards of living while at the same time reinforcing the traditional cosmopolitan trading function, which was recognized in 1958 by the creation of Manama as a free transit port.

The environmental impact

One advantage the miners in the arid lands have over their compatriots operating in the humid lands is that their environmental impact is less likely to bring rapid outraged responses from neighbouring resource users. With minimal population densities, there are not too many neighbours, and with a highly paid but often temporary workforce, out to make as much money in as short a time as possible for families often many thousands of kilometres away, there is little likelihood of any concern by the workforce for the effects of mining upon the immediately adjacent neighbourhood.

Yet the size of the mining areas can be considerable and the effects of air pollution, for example from copper and lead smelters, can affect vegetation for 10–20 km downwind. Where mining is taking place in the developing countries the constraints in the way of legal controls on pollution are likely to be less. One group of researchers suggests that for unscrupulous developers anxious to keep their costs down (and pollution control involves considerable extra capital costs) the 'freedom to pollute' in such countries is a large incentive (Frobel, Hendricks, and Kruge 1979).

Pollution has been only part of the impact. The beneficial spin-offs of geological knowledge (particularly of water supplies) and improved accessibility from the mining roads, railways, and aircraft landing strips have to be offset against the removal of timber and any combustible vegetation for many kilometres around the mine site and the associated problems of increased soil erosion. In the case of Broken Hill, Australia, the summer dust-storms became so bad in the 1930s that the mining companies provided funds to the town council to plant protective tree belts around the town from 1936 onwards. By the 1970s a 'green belt' of native trees and shrubs had been established successfully around the town and was partly in use as a recreational area (Mabbutt 1978).

The future of mining

If past evidence is any guide, the increased knowledge of arid lands geology resulting from the oil searches will bring the inferred and hypothetical mineral *resources* into the category of *reserves*, particularly as prices for the ores continue to rise (see Fig. 9.2). The demand for base metals such as iron ore is forecast to increase markedly over the next few decades and there is no doubt that more arid land iron ores will find their way on to the world market.

Economies of scale

To the extent that the production costs of the arid lands are significantly higher than elsewhere, the usual response of developers, whether governments or private companies, is to aim at economies of scale. The result, therefore, is likely to be concentration of production upon a few very large projects, and Skinner's model of mining trends may well apply (Fig. 14.2). For the older established mining areas, consolidation of mines may still lead

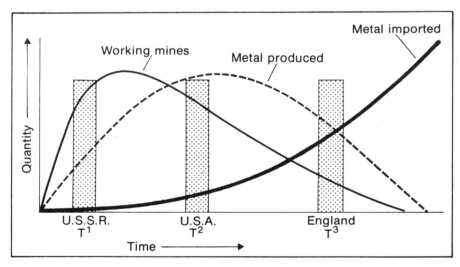

Fig. 14.2 Model of national mining production sequence. *Source*: Skinner 1976.
   Notes: The sequence of development of number of working mines, production and metal imported are suggested for three stages of national development (T_1, T_2 and T_3) with a nation identified at each stage.

224

to increased production but with a smaller labour force (point T_2 on Fig. 14.2). For most arid nations, however, the current situation is probably closer to point T_1, with the number of mines increasing and production expanding commensurately.

Given the high cost of mining development it is likely that private ventures will draw more and more upon international funding with the resultant return flows of a share of the profits, if any, out of the arid nation to the international shareholders. The mere size of such international concerns is itself a source of concern to some commentators.

The mining ethic

Lewis Mumford in his book *The Pentagon of Power* noted that mining introduced some questionable concepts into resource-management strategies:

> [It] set the pattern for later modes of mechanization by its callous disregard for human factors, by its indifference to the pollution and destruction of the neighbouring environment, by its concentration upon the physico-chemical process for obtaining the desired metal or fuel, and above all by its topographic and mental isolation from the organic world of the farmer and craftsman and the spiritual world of the Church, the University and the City. In its destruction of the environment and its indifference to the risks to human life, mining closely resembles warfare.   (Mumford, quoted in Coombs, 1979: 9)

Thus mining, in the hands of the bureaucratic machines of the international corporations, which exercise power without personal responsibility 'but rather vicariously and impersonally on behalf of the corporation – an entity without compassion to be moved or soul to be damned' (Coombs, 1979: 9), has the potential to exploit not only the environmental but also the human resources of the arid lands. When the budget finances of such mining companies greatly exceed those of the underdeveloped arid nations in which they operate, then we must ask who is really planning the national resource development?

CHAPTER 15

Urban centres

Given the low productivity per unit area, the limited population densities, the lack of water, and poor timber supplies for construction, it is surprising to find large cities in the arid lands. Nevertheless, by the early 1970s at least 89 cities of over 100 000 population were spread through the arid lands, 37 of them on the coasts. Nine (Cairo, Karachi, Tehran, Lima, Alexandria, Baghdad, Lahore, Baker, and Phoenix in descending order of size) had over 900 000 inhabitants (Wilson 1973). Some, such as Cairo, can be counted as some of the first cities, others such as Tehran (founded 1788) and Phoenix (incorporated 1881) are more recent additions. Their origins and functions illustrate the variety of factors which have led to their creation and the extent of their success in offering living spaces buffered against an adverse environment.

General influences on arid urban centres

Arid space

Because so much of the arid lands is either unproductive or provides a very low return per unit area, it is not surprising to find that urban centres are few and far between. Where they do occur, they combine many functions which in more productive humid environments would be spread between several complementary centres – perhaps as a hierarchy of functions and services. Theories of the spacing of urban centres evolved in more evenly productive humid environments, such as Christaller's Central Place Hierarchy developed in the pre-motor-age 1930s in southern Germany (Christaller 1966). Such theories can be applied only to those portions of the arid lands where productivity is sufficient to support a sedentary and evenly spaced agricultural population. Such areas would be the richer rain-fed and irrigated agricultural lands, provided the settlement had been established

226

long enough to have allowed the historical sorting effects of travel costs and service provision to have evolved without major external planning.

The patterns of urban centres reflect varying approaches to the management of arid space. In many cases the urban centres serve only their own population and have only tenuous links with the countryside around them. The traditional 'caravanserai' overnight accommodation centres on the caravan routes across the deserts, the mining towns, the military posts (guarding the agricultural from the nomadic populations, or as regional military headquarters or, more recently, as test sites for weaponry), all had very limited relationships with such local population as did exist. On the Australian arid zone military rocket range at Woomera, at the peak of its activity in the late 1950s a town of 4500 people (designed largely on English New Town plans of the 1950s!) offered basic retailing, entertainment, and health welfare services. The surrounding ranching population which previously had obtained supplies from Port Augusta (175 km away), were provided with special security passes to shop in Woomera on Saturdays, but it was not *their* custom which supported the town.

The Israeli port of Eilat was established in 1949 on purely strategic grounds, as the only possible port for Israel with access to the Indian Ocean and the Far East. At the time of its establishment its site offered:

nothing to attract human settlement: no water, no soil, no impressive
mineral wealth, no centrality of location.   (Karmon, 1976: 106).

Deep water close inshore enabled the subsequent establishment of an oil-tanker terminal, but the foundation and support for the town, which by 1975 had 15 000 inhabitants and claimed to be a dormitory for the copper mines at Timna as well as a port and tourist resort centre with 370 000 visitors per year, was a political decision. Once the copper mine was closed in 1975, the survival of the town depended upon subsidies from the national treasury (Karmon 1976).

Arid sites

The detailed location of cities in the arid lands often shows remarkable continuity which may be a reflection of the limited possible sites. Thus, one of the oldest continuously inhabited urban sites is Jericho, where springs in the alluvial fans along the sides of the Great Rift Valley provided the initial and one of the continuing factors supporting sedentary occupation. Throughout North Africa, as well as southwest and central Asia, the alluvial fan has been a favoured site for cities, since in many cases the city has grown from villages using the water and soil of the site as the basis for irrigated agriculture. In the layout of these cities the original field patterns still show in street and house alignments, as we shall see.

The long history of human occupation of the semi-arid fringes of the Fertile Crescent from Egypt to Mesopotamia is reflected in the thousands of

227

ancient settlement sites. In many cases the original sites are marked by 'tells' – the accumulated debris of past settlements which now form man-made hills several metres high, alongside which the modern settlements are still located. From this archaeological evidence continuity of human occupation for 10 000 years has been documented at Jericho.

The environmental stresses and the need for shelter

In terms of the critical temperatures for a naked person at rest the comfort zone seems to be a range from 10 °C to 27 °C. Whether there are inherent racial differences in tolerances is debatable, since the evidence of different settings for air-conditioning equipment around the world may reflect social conventions and heating/cooling costs. Furthermore, the work of researchers such as D. H. K. Lee and W. V. Macfarlane has demonstrated that careful control of activity and diet will allow human activity to continue under a wide range of aridity stresses.

The stresses of, and associated antidotes for, arid climates fall into four groups:

1. Excessive heat load on the human body, reduced by shading from insolation and increased sweating.
2. Desiccation, prevented by adequate water intake plus salts to allow water to be absorbed. Water losses through sweating of 1 litre h^{-1} without

Table 15.1 Design principles for housing and clothing in arid lands

Objective	Principles for housing	Principles for clothing
Reduce heat production	Minimize heat and vapour production from cooking etc. ←———————— Functional	Light weight Unrestricted convenience ———→
Reduce radiation gain and promote loss	Minimize solar insolation High reflection of shorter wavelengths and emission of longer wavelengths Convection over surfaces heated by radiation ←——— Insulation——————————→	Shade Reflection
Reduce conduction gain	←——— Insulation ——————————→ ←——— Wind exclusion————————→ ←——— Air cooling————————————→	
Promote evaporation	←——— Wetting ——————————————→ ←——— Maintain convection ————→ over evaporation surfaces	

Source: Lee 1969.

replacement would cause 5 per cent reduction of body weight in a 68 kg man within 3 hours; a reduction to 10–12 per cent would be fatal.

3. Excessive skin exposure to sunlight, resulting not only in the above factors, but also in photo-chemical lesions (blisters, dermatoses, acne) and possible skin cancers. A separate but related problem is eye-strain and possible blindness from the reflected glare of light-coloured ground surfaces under high sunlight.

4. Psychological stresses – both indirect from the isolation and lonely life of some arid land resource-use systems, and from the effects of the hot dry winds. The desiccating effects of such winds together with the high static electricity associated with them apparently has deleterious effects on humans. Scientific documentation of such effects is difficult to find, but the occurrence of the local hot desert wind the 'Santa Ana' (with relative humidities from 5 per cent to 20 per cent and speeds of 24–80 kmh^{-1}) in Los Angeles has been claimed to be associated with above-average numbers of homicides (Miller 1968).

The design principles of shelters to protect against the three physiological stresses above are known (Table 15.1). Most have been consciously applied in traditional building styles in the arid lands. However, the internationalization of building styles around the world from about the 1950s onwards has posed problems which have been solved only partially by the application of air-conditioning systems. The fourth, psychological, stress has been recognized relatively recently and coping strategies are sparsely documented.

Urban designs

The casual traveller to the arid lands of Africa and Asia rapidly comes to the conclusion that two basic styles dominate the urban areas. On the one hand are the narrow, winding streets of the older traditional pre-nineteenth-century portions of the cities, cluttered by shops and pavement stalls, thronged with pedestrians, and bounded by the heavy, compact, introspective structures of the houses, markets/bazaars, mosques, temples, and palaces (Pl. 15.1). On the other hand are the broad, straight, sun-soaked streets of the 'new' towns with the latest international-style hotel and office blocks rising above the sprawled open-style modern housing in which the air-conditioners hum and from which the cars issue forth to carry the inhabitants along the relatively deserted streets to the office, the mosque, or new shopping centre.

In the arid lands of the Americas, southern Africa, and Australia the impression is different. Here it is the 'traditional' older centres *as well* as the modern additions which have the geometric street grids showing their planned origins. Whether the grid of the Spanish colonial city centred on the church, plaza, and administrative headquarters in Central and South Amer-

Plate 15.1 The Jewellery Bazaar, Bukhara, USSR, photographed 1976. The traditional vaulted-dome structure covered four major 'streets' of shops leading to the central meeting-place under the largest dome. In effect the whole floor and street space is shaded from direct sunlight. The current commercial function, however, is limited. At left is Minaret Kalyan (twelfth century) and the rear of one of the sixteenth-century university complexes.

ica, or the grid of the speculators or railway surveyors in the arid lands of North America, or the government and privately surveyed grids of southern Africa and Australia, or the layout of the collective farms of central Asia (Pl. 15.2), the patterns have the same basic components of rectangularity. Peripheral to this grid of streets and single houses may be a 'shanty-town' of motley dwellings with winding narrow alleyways, housing groups of itinerant Aborigines in Australia or blacks in southern Africa.

Between the traditions of the African and Asian designs and the 'modern' styles which have begun to oust them and which already had been introduced into the arid lands of the Americas, southern Africa and Australia during the nineteenth century, there is a fundamental difference in the effects, even if not the intent, of the urban design. The apparently 'cramped' urban spaces of the traditional African and Asian cities provided large areas of shade for both streets and house walls and considerably reduced heat loads. By contrast the broad street grids of the 'modern' styles, together with the low-density independent housing structures, with few, if any, shared walls, meant that protection from solar heating was minimal, temperatures inside the buildings were higher and had to be controlled by air-conditioning.

The spread of Western (i.e. European and eastern North American) architectural styles into those oil-rich arid nations of southwest Asia and northern Africa which could afford to pay for it introduced alien designs and materials (particularly steel and glass) basically unsuited to arid conditions. Yet the cheapness of thermal electricity based upon cheap oil provided the cosmetic of air-conditioning from the 1950s to the 1960s to more and more of the public and private buildings in these nations. As the cost of oil rises, however, so does the cost of maintaining these air-conditioned environments and the folly of their designers and owners becomes more apparent. Indeed, one effect of the 'energy crisis' of the 1970s may be to the benefit of the arid nations in that it may encourage the design of structures more appropriate to the environmental stresses of the arid lands and less dependent upon fossil fuel energy for maintenance of a comfortable interior environment.

Traditional designs

Living spaces
Traditional housing in the arid lands used local materials, whether earth, mud-brick, fired brick, stone, or timber, to buffer the occupants against heat and dust inputs by thick walls and few windows. To benefit from night-time convective and radiative cooling roofs were used as bedrooms (Pl. 15.3) and roof 'wind catchers' funnelled cool night air into the rooms below (Bourgeois 1980). The courtyard, often with central water basin, encouraged convective cooling and seems to have been a basic design principle from Mohenjo-Daro

Plate 15.2 New irrigation settlement, central Asia, USSR, photographed 1976. State farm No. 5 village, Hungry Steppe, near Gulistan, established 1961, with standard semi-detached housing, electric power, television aerials, and the air of a pioneer settlement.

231

Plate 15.3 Street in Old Kokand, USSR, photographed 1976.
Traditional walling construction at left, using dried mud with timber beams plastered over, and sleeping areas under shade house on roof – simple, cheap, and effective.

in the Indus Valley to the Moorish architecture in Spain, from the caravan-serai to the private house in Asia (Fig. 15.1). Where seasonal temperature contrasts were considerable, enclosed well insulated winter quarters contrasted with shady but well ventilated summer quarters. In the Iranian cities, one side of the house courtyard would be roofed but open to the court-yard – the *iwan*, while an underground room below it could be used in the extremely hot season. The remaining enclosed buildings around the court-yard were used in winter (Bonine 1979). Contemporary 'folk' architecture still uses cheap but effective designs as witness the Indian 'ramada' of the southwest USA (Pl. 15.4).

Urban layout
There is considerable debate about the geometries of traditional city designs in the arid lands, debate which has spilled over from general debates on the global origins of cities. Various types of cities have been suggested including a 'Muslim City' (Grunebaum 1955), a 'Middle Eastern City' (Lapidus 1969), and a 'Black African City' (Winters 1977). Elements of all three of these seem to be present in the African and Asian arid lands.

According to Grunebaum, the Muslim City was dominated by two central institutions – the mosque and market; the division of residential space into distinct, often ethnically based, areas; a hierarchy of trades (from most important to least important) out from the central mosque; and a surround-ing maze of narrow twisting streets. Lack of centralized overall control of urban development was seen as the main factor responsible for this apparently haphazard development.

The Middle Eastern City of Lapidus and colleagues was a variant of the Muslim City but tied more directly to the country both in its origins and contemporary functions. The city was seen as a composite of several cities rather than one single entity. The ethnic groupings or quarters of the cities,

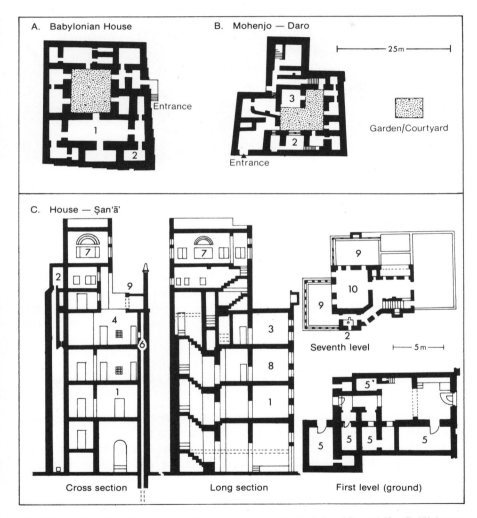

Fig. 15.1 Traditional house designs. *Sources*: A, B Schneider 1963; C Kirkman 1976
Key: for all diagrams 1 – Main living room; 2 – Toilet/bathroom; 3 – Kitchen; 4 – Court (*sanā*); 5 – Animal stalls; 6 – Well; 7 – Recreation room (*mafraq*); 8 – Reception room (*diwan*); 9 – Terrace; 10 – Women's quarters

Plate 15.4 A 'Ramada', Papago Indian Reservation, Arizona, USA, photographed 1969.
Made from scrap timber and corrugated iron sheets this formed an effective shadehouse and eating area in temperature over 45 °C. The low wall reduces ground heat radiation gain and keeps out some dust, while the open sections encourage through-ventilation cooling.

such as in Baghdad, Kufra, and Basra, were traced back to traditional bedouin seasonal camp sites. Within the cities was much open space, as courtyards, gardens, orchards and fields. They suggest that the city was less a built-up area than a regional complex surrounded and in some cases internally divided by walls. As an example

> the name 'Bukhara' applied both to the city proper and to the oasis region as a whole. Such oases may be regarded as extended boundary, multiple settlement composites. (Lapidus 1969: 68)

The city had, as with Qum, Merv, and Kazerum, grown from the enclosure of clusters of villages and the markets which served them.

The Black African City noted by Winters was even more open in its design with predominantly single-storey structures clustered in walled or fenced compounds around the ruler's palace. Streets were irregular and apparently unplanned.

The arid land cities of Africa and Asia seem to contain elements of all three – reflecting the complex interplay of cultural and environmental constraints in their design. The complex pattern of land use within the *contemporary* urban areas, however, reflects the gradual expansion of the built-up area over time, slowly engulfing the surrounding fields and villages (Pl. 15.5). In Iran there is specific evidence to show that the apparently irrational contemporary maze of side-streets and cul-de-sacs directly reflects the prior alignment of irrigation canals (which when dry *were* the streets) and rectangular fields. Indeed, in this preservation of prior agricultural landscape patterns in the contemporary urban fabric, the arid land cities are no different from the cities of Europe or of the humid lands of Asia (Bonine 1979; Ward 1964; Wheatley 1971).

Within the complex of streets, however, certain distinctive features did dominate the layout. The minarets of the central mosque, which is often oriented to the direction of Mecca and which may have affected the immediate street pattern around it, will be the tallest structures, while the ruler's palace may be the largest in ground plan. In addition, the bazaar may be a street of speciality shops or may be located in a specific building – as the Jewellery Bazaar in Bukhara (Pl. 15.1) which anticipated the enclosed shopping arcades of the late nineteenth century in Europe and the modern enclosed shopping centres.

In detail the city layouts will vary, showing the influence of topography, the presence or absence of reticulated water systems making for greater or lesser obvious street design, and the height of buildings reflecting population pressures, building materials, and potentials for urban sprawl. The earlier dismissal of the 'maze' of streets as unplanned, however, must now be modified and the patterns recognized as having evolved from complex demands

Plate 15.5 New flats, Fergana, USSR, photographed 1976.
Little concession made to local climate apart from green 'buffer' of vegetation of 'garden' area. Note lack of window shade protection. Relatively cheap but probably not very comfortable housing in 'heat wave' conditions.

upon, and human responses to, the constraints of the site. Any attempts to modify the layout of such traditional areas of the city in future will have to bear such demands and responses in mind.

Modern designs

The creation of new urban centres or of extensions to traditional centres in the arid lands since the late nineteenth century has been influenced by principles of design which owe little to the traditional arid cities. Many centres were created as part of the invasions of the arid lands from Europe, where the basic principles of land allocation already had been decided. In some cases, particularly the mining towns, the humid area designs were deliberately retained to be a familiar element for the immigrant workforce coming from the humid lands into an otherwise unfamiliar and inherently hostile environment. Such designs, although psychologically attractive, were only habitable with the provision of constant air conditioning.

Living spaces

Apart from the problems of the transfer into the arid lands of the humid area housing designs (Marshall 1963; Saini 1970), there is some evidence of attempts to design living spaces which apply the basic principles listed in Table 15.1. One fundamental continuity in design of living space is the use of the courtyard. Commenting on the antiquity of the concept, an Australian architect (Saini 1962: 4) suggested:

> This widespread use of the court is to enclose not only the interior living spaces but to frame a part of the sky, to pave a few yards of desert and to invest these fragments of nature with man's mark, a symbol of shelter. Whatever its emotional origins may be the courtyard, with its infinite possibilities for variations, still provides a satisfactory answer to a number of modern design problems.

He went on to illustrate how the courtyard principle was used in the Negev by Israeli architects to stimulate cooling air flows where winds are relatively few, and in Morocco, Iraq, and India from even the smallest housing units to the design of communal open space (Fig. 15.2).

Paralleling these attempts at low-cost building design are the air-conditioned humid designs set down in the arid lands. Contrasting types of such housing styles in northern Australia are shown on Table 15.2, where their relative costs in the 1960s are compared. For private housing, the costs of air-conditioning are paid for by the mining companies and must be set against the income from the mining operations. In contrast, the government housing relies upon structural design to reduce heat loads and appears to have been significantly cheaper both to construct and to maintain.

Urban layout

The dominance of the planned town – planned in the sense of the individual

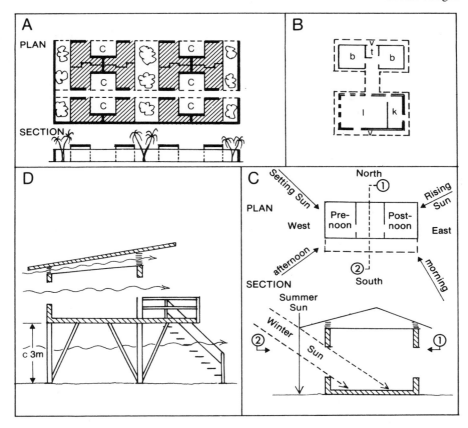

Fig. 15.2 Modern house designs. *Sources*: A, B Saini 1962; C, and D Golanyi 1979
Key:
(A)   Housing in the Negev, Israel
Interiors shaded, C – courtyards.
(B)   House with both heavy and light wall structures. The heavy walls
(double lines) provide day-time living space because of the time lag in
interior heat build-up, while the light walls (single lines) provide rapidly
cooling areas for sleeping at night
l – Living area; k – Kitchen; v – Verandah; b – Bedrooms; t – Toilet.
(C) Plan and section of simple design for reducing heat loads. As with B
the area used varies according to time of day, in this case influenced by
solar aspect (in the northern hemisphere) and the need in some areas for
winter warming as well as summer cooling
(D) Design for reduction of ground heat gain and maximum use of
through ventilation
Arrows show air movement through the structure, which being at least
3 m off the ground, avoids much of the ground infra-red heat radiation

237

Table 15.2 Housing types in northern Australia and climatic control costs, late 1960s

Characteristics	Government housing	Private housing	
		(a)	(b)
Building materials	Timber frame; asbestos sheeting; corrugated iron roof; raised on stilts	Brick veneer; tile roof	Timber frame; asbestos sheeting Portable
Floor space (sq. ft)	1100	1600	1000–1450
Building cost per sq. ft	$12–17	$20	$26–28
Total house cost	$13 000–14 000	$32 000–34 000	$25 000–40 000
Climatic controls	Ceiling fans; eaves; louvres; insulated ceilings and walls; east–west alignment.	Air-conditioned throughout	
Weekly economic rent (returning 8% on capital)	$25	⟵ $96 ⟶	
Actual weekly rent	$17–18	⟵ $6 ⟶	

Source: Roberts and Sheridan 1969.

238

blocks and street layouts existing on paper before they were imprinted on the ground – is widely evident in modern centres. On the semi-arid prairies of western Canada the settlement pattern mirrored plans on the Great Plains, portions of southern Africa, and Australia:

> The elements of the plan were the grid survey system, the vast rail network, the dispersed farmstead and the trade centre. (Rees, 1969: 30)

Since they were founded as part of the grand design for the agricultural occupation of the land, their sites could be regulated, 8–16 km apart along the railway lines, with larger centres 176–208 km apart (where train engines and crews were changed). Even their names were dictated by the bizarre whims of the planners, as illustrated by the A–Z sequence of towns and rail sidings in Saskatchewan from Alma to Zeneta! In Australia the surveyors not only demarcated the agricultural holding boundaries but also chose and laid out the potential town sites; in South Australia from the 1860s onwards as mirror images of the capital city's design – an urban centre with surrounding parklands and beyond the suburban plots.

The resettlement of nomadic groups in central and southwest Asia and North Africa has required that this basic geometry, introduced from the humid lands, be adapted to traditional spatial organization. Such new settlements in Abu Dhabi in the mid-1970s showed some concessions to traditional settlements in the layout of tribal and clan groupings, livestock and cultural services (Fig. 15.3).

Urban functions

The functions of the urban centres in the arid lands are many and varied. The evolution of the cities in Mesopotamia resulted from the grafting on to village functions (dormitory for the rural workforce and low-order service provision) of the functions performed by the 'non-productive' population of soldiers, priests, and bureaucrats, supported by the food surpluses of the irrigated lands. A Sumerian inscription of 2650 BC noted that the King:

> built the temple and the walls, erected statues, dug a canal, and filled the storehouse with grain. (Schneider 1963: 35)

The city as security from hunger and attack and sanctuary of gods and rulers was in the making.

The multiple functions of the arid cities combine most of those found in cities beyond the arid zone. As Schneider (1963: 14) has suggested, 'Babylon is everywhere':

> In Babylon was assembled, for the first time, everything that constitutes the attractiveness and the danger of giant cities: culture and depravity, arrogance and money, temples of faith and those of hectic amusement, splendour and misery.

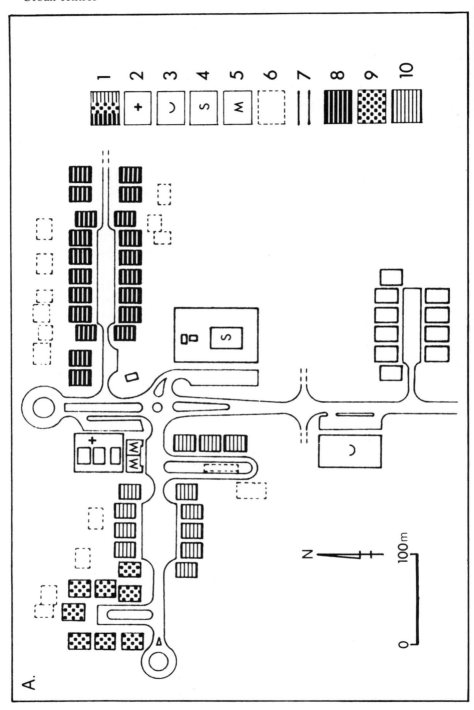

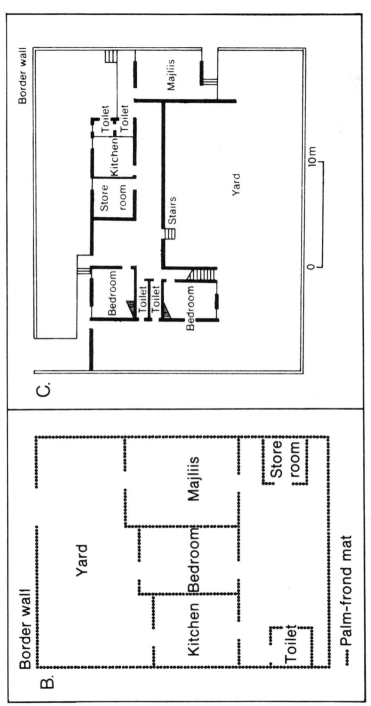

Fig. 15.3 Traditional and modern settlements in Abu Dhabi. *Source*: Cordes and Scholz 1980
(A) *New settlement of Al-Wathba in 1978 (established 1972)*
1 – Low-cost housing. Shading indicates tribal groups (see 8, 9 and 10), unshaded = not yet allocated; 2 – Clinic; 3 – Mosque; 4 – School; 5 – Market (Souk); 6 – Livestock compound; 7 – Roads; 8 – Al bu Shar tribe housing; 9 – Al bu Rahmah tribe housing; 10 – Al bu Mundhit tribe housing
(B) *Traditional hut (Barasti), Abu Dhabi*. *Majliis* – general living area. Walls are of palm-frond matting. Scale as for C.
(C) *New low-cost house, Abu Dhabi, 1978*

Thus the markets, the religious centres – mosques, cathedrals, prophets' tombs, and religious colleges – the palaces and the barracks are common components.

In detail, however, these functions have changed over time. There is some evidence that the cities in the arid lands formed not only the earliest urban areas as Schneider implied above, but also contained the largest proportion of the national populations before the Industrial Revolution concentrated population in the burgeoning cities of Europe from the late eighteenth century onwards. About 1800, according to Lapidus (1969), the most urbanized nations in the world were the arid lands of southwest Asia with 10–15 per cent of their population in urban centres compared with European figures of 1–7 per cent. Only with the spread of industrialization and steam transport did urbanization in Europe catch up with and eventually overtake the Middle East. The Europeanization of the various southwest Asian national economies, particularly since the 1920s, however, has renewed the growth of the urban populations. For example, Egypt's urban population grew from 21 per cent in 1917, to 24 per cent in 1937 and 40 per cent in 1966 (Issawi 1969). Such changes brought modifications in both form and function of the cities and towns.

Changing oases in the Sahara

By the 1960s there was evidence both of expansion and decline in the oases around the edge of the Sahara. Of the ten oases with over 10 000 population, seven were in Algeria (Fig. 15.4). Several of the oases were losing population as a result of removal of traditional functions and outmigration of workers. Murzuk, sometime capital of the Fezzan, had 10 000 inhabitants in 1900 but only 3800 in 1964, having lost the traditional caravanserai function and workers to the oilfields in the north of Libya. Nearby Sebha had benefited after Libyan independence in 1951 from the establishment of the new regional administrative headquarters and expanded airport facilities, and was growing rapidly – in 1964 having 10 000 population.

In Morocco, Figuig with 12 000 population had lost workers to France, its palm gardens were neglected and trades and services declining. Atar in Mauritania with 6000 population was benefiting from nomad resettlement around the original (1909) fort and oasis. Ouargla with a population of 20 000 was the traditional summer camp for the Mekhadma nomads (some 5000 strong) after seasonal movements of up to 400 km. Expansion of the wells by new bores after the 1950s, plus opportunities in the nearby oilfields, had encouraged the nomads to become farmers, labourers, and traders. By the mid-1960s only 5 per cent of the tribe were still practising nomadism.

A schismatic Islamic sect (Kharijites) had established a village in the late nineteenth century at Ghardaia with central mosque and concentric circular street pattern. In 1896 a population of 8300 was based on the oasis. By the mid-1960s the population had grown to 30 000 and they had become known

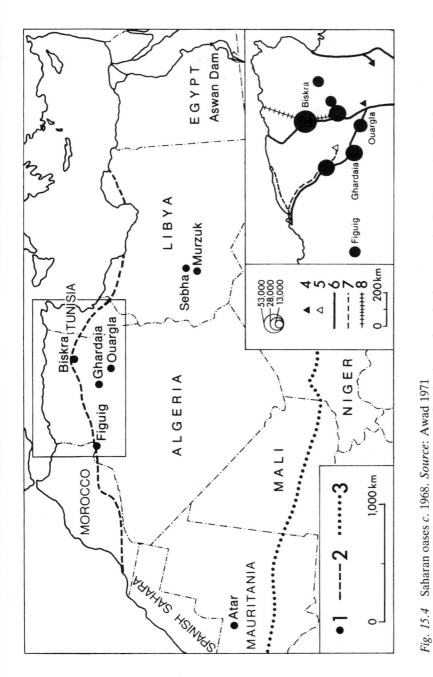

Fig. 15.4 Saharan oases *c.* 1968. *Source:* Awad 1971
Key: 1 – Oasis towns noted in text; 2 – Northern edge of palm groves; 3 – Southern edge of *had* grass;
Inset Proportional circles indicate town populations
4 – Oilfield; 5 – Gas field; 6 – Oil pipeline; 7 – Gas pipeline; 8 – Railways

as the grocers of Algeria – running grocery businesses by motor truck and bus services throughout the State. The town's oases gardens had become not the basis for its existence but country 'retreats' to retire to or commute from.

The largest oasis, Biskra, with 53 000 inhabitants, had been a fortress first for the Romans, then the Turks and French: then an agricultural centre, a market centre for adjacent oases and nomads; a caravanserai on the edge of the desert; an administrative headquarters; French winter tourist resort, and finally an oil supply depot.

The new oases of central Asia

As part of the Russian plans for the development of the Turkmen region of central Asia, a series of irrigated state farms have been located along the northern banks of the Kara Kum Canal. After a massive earth-moving project in which the bajadas sloping north from the Kopet Dag Mountains were graded, roads built, and new towns created as the diverted waters of the Amu Darya River were gradually extended westwards, the nomad Turkmen and Uzbek peoples were resettled as agriculturalists producing the national cotton crop and vegetables (especially water-melons). Within the settlements yurts and camels remain as relics of past life styles, the one still used as accommodation alongside the standard brick housing units but the other now serving only a sentimental function as a link with the past (see Pl. 10.1).

Ashkhabad had been reached by the Kara Kum Canal in 1962 and the state farm 'Way of Lenin' had been laid out in 1968 on 9716 ha. Originally 4800 ha had been irrigated but by 1976 some 7110 ha were being watered. The settlement was mechanized, its 496 agricultural workers using 272 tractors, as well as mechanical cotton pickers and motor trucks, and possessing 134 cars and 106 cycles. Wages were three and a half times as high as urban wages elsewhere in Russia, as an incentive to stay on in this pioneer settlement. However, large families (averaging 7–8 children) posed problems for future employment since the size of the settlement – some 5400 people – would offer no alternatives to the agriculturally oriented employment.

As one of the series of the new settlements along the canal the State Farm was contributing to the central Asian region's dominance of the USSR's cotton production (93%) and cotton-picking equipment (100%) as well as silk and karakul pelt production (66%) (Akramov 1976). In this case the region's irrigation potential was being used to meet the national demand for subtropical crops and raw materials – from silk to cotton, from rice to watermelons.

Tourist centres: old and new

Tourism – travel for pleasure – if broadly enough defined has been a remarkably ancient activity associated with the urban centres of the arid lands. Broadly interpreted, tourism might be classified as for either religious, medical, educational, or entertainment purposes. Pilgrims travel to holy places as part of their religious duties and social needs. The sick travel to specific environments to take the air or the waters at the spas, or be cured by religious faith. The curious travel to find new experiences and to authenticate cultural history. Finally, the recreationists seek to be entertained either by their own exertions or the activities of others.

Religious movements to holy places in the arid lands include the pilgrimages of the Christians to Jerusalem and Muslims to Mecca, and a host of other places associated with historic events, birthplaces or tombs of saints (Pl. 1.2). The British travel firm, Thomas Cook, ran its first excursion of fifty tourists to Jerusalem and the Holy Land from Britain in 1869, but they were following in the footsteps of European Christians and Crusaders from at least medieval times. The modern Muslim pilgrims, journeying to Mecca by jumbo jet, are maintaining 1300 years of annual migration to what was originally merely a sixth-century caravanserai on the route from the Yemen to southern Jordan and Palestine (King 1972).

Medical movements to the arid lands have been associated mainly with the belief from the nineteenth century onwards that tuberculosis could be cured by the dry air and high ultra-violet radiation of the deserts. After beginning as an overnight stopping point on the California – east coast stage coach route, Tucson in Arizona benefited from a rail link in the 1880s and was advertised as a successful TB sanatorium at the turn of the century. It is still recommended as a site for sufferers from lung ailments and asthma, but now gets support from the local film-making industry, an air force base and university as well as general recreational tourism (see Pl. 1.4) (Wilson 1980).

Educational movements to the arid lands have been of two types, the one to marvel at the landforms, fauna, and flora, the other to examine the evidence of past civilizations preserved in the arid lands. Whether the saharan sand seas, the canyons of the Colorado River (USA) or the Fish River (Namibia), the badlands of the Great Plains (USA), or Ayers Rock (Australia), the colours, textures, and dramatic forms of the land have an aesthetic appeal which an increasing number of tourists find attractive. At the same time the Great Pyramids of Egypt, the tells of Mesopotamia, the Great Wall of China, the cave palaces of Mesa Verde (USA) and the ruined cities of central Asia provide the sense of history which permeates the arid lands.

The recreationalists combine some of the above activities with the major motive of entertainment, whether hiking, boating, pony-trekking, gambling, or attending theatrical performances. *Their* city above all others is Las Vegas.

Las Vegas: unique arid city

In 1970s a population of 275 000 occupied a sprawling urban area in the Nevada Desert with an average rainfall of 97 mm and desert scrub vegetation. Artesian water had drawn Mormon settlers in the 1850s to the limited agricultural potentials, but by 1910 the population was only 5000, although a railway junction had been established in 1905 and some mining of gold and precious metals had taken place locally. By 1920 mining had benefited from the First World War boom and the population reached 9000, but the Depression of the 1930s and the exhaustion of the ore bodies slowed growth and in 1930 the population was only 10 000.

In 1931, however, the State of Nevada passed the 'Liberal Laws' which made gambling legal, allowed expedient marriage (by reducing the prior residence requirements to only 6 weeks) and expedient divorce (by speeding up and simplifying the legal procedures). Paralleling this legal development came the construction of the Boulder (now Hoover) Dam on the Colorado River. As a result by 1934 cheap hydroelectric power was available, together with a new source of water (Murphy 1971).

The Second World War added a further dimension to urban development when, in 1940, the Las Vegas Bombing and Gunnery Range and Nellis Air Force Base were established immediately north of the town, which became a dormitory for civilian and some military personnel. By 1940 the population was 15 000, but by 1950 this had risen to 50 000 and the Nevada Atomic Bomb Test Site had been established. During the 1950s massive expansion of air-conditioned hotel accommodation specializing in high-quality live entertainments for the ever-increasing tourists (gamblers, honeymooners, divorcees, and boat enthusiasts operating on Lake Mead) took place, along with the expansion of local electronic industries, partly serving the military establishments.

From the 1960s onwards the tourist boom has continued. The city population reached 130 000 in 1960 and 250 000 by 1965. Freeways enabled Los Angeles patrons to drive up for weekend recreation and frequent air services linked Las Vegas with the major American cities.

The desert basin had been re-evalued several times between 1900 and 1980. The limited agricultural resources soon had been recognized, but the main success in terms of population growth has come partly from the empty space of the desert basin and the use of this for military purposes – both the clear atmosphere for flight training and the 'empty' land for weapons and equipment testing. The success also has come partly from the unique legal situation by which the national code of ethics and morals is modified in this particular state. As the two largest towns in the State at the time of the legal changes, Reno and Las Vegas were quick to benefit. Thus, with 50 per cent of the workforce directly or indirectly employed in the gambling and recreation industries, 18 per cent in military jobs, and 29 000 marriages per year compared with 900 in an equivalent sized town elsewhere in the USA, Las Vegas is a very special arid lands city. The risks of life here are not only confined to the slot-machines or card tables. In 1980 an underground nuclear test leaked radioactive materials into the atmosphere.

Tourist impacts: boon or bane?

The major tourist impact on the urban centre is economic. Direct payments for accommodation, goods and services, visas and entry permits provide income while employment in service industries generates extra cash flows. Clawson (1963) estimated that a quarter of the tourists' dollars accrued to the tourist area, while Australian studies have suggested that one tourist dollar generated $3.40 worth of business to the community as a whole. A Meccan saying is:

> We sow no wheat nor sorghum, the pilgrims are our crops.

(King 1972: 62)

With over 400 000 pilgrims in 1970 and over 2 million in 1980, each paying a $12 entrance fee and bringing approximately $200 for expenses, the 'harvest' was bountiful. For Israel the 750 000 annual tourists in the early 1970s were a similar source of income.

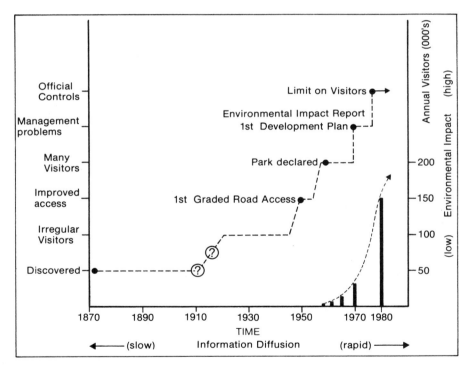

Fig. 15.5 Tourist use and environmental impact at Uluru (Ayers Rock–Mt Olgas) National Park, Australia. *Source*: Ovington *et al.* 1973 and various
Note: The increase in absolute numbers of visitors (1970–80) and associated increased impact resulted from a combination of increasing information and improved access, which in turn led to management problems

247

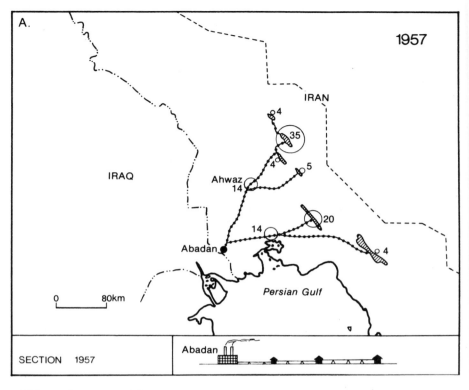

The environmental impacts, however, can be considerable and less beneficial. The vast numbers of tourists moving into and through the arid lands bring health and quarantine risks, and accommodation and services may be used for only brief periods of time. Thus the Muslim pilgrims' camps at Jed-

Table 15.3 Multiple management goals for Ayers Rock–Mt Olgas National Park, Australia, 1958

Goals

1. To preserve the outstanding scenic and geologic features of Ayers Rock and the Olgas.

2. To preserve a reference area of Australia's arid zone ecosystem.

3. To preserve sites of Aboriginal cultural significance, including the integrity of these sites.

4. To provide controlled use of the Park's resources for the purpose of:
   (a) appropriate recreational activities,
   (b) education,
   (c) scientific study,
   (d) wilderness experience.

Source: Ovington *et al.* 1973.

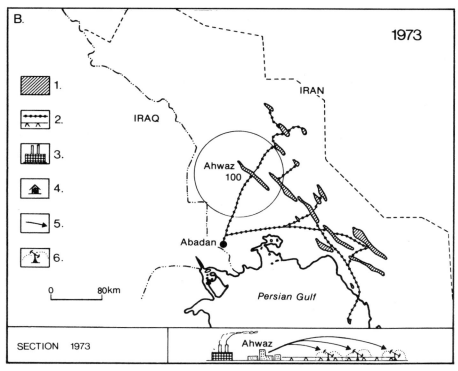

Fig. 15.6 Urban satellization in Iran. *Source*: Melamid 1973
Key: (A) 1957 situation; (B) 1973 situation (forecast)
For each period a map of oilfields and urban centres and a section from
the oil refinery at Abadan to the oilfields showing communication systems
are provided. Dashed line delimits company concession area and numbers
show percentage of employees in urban centres
1 – Oilfields; 2 – Oil pipelines; 3 – Oil refinery (Abadan); 4 – Urban
settlements, 1957 section only; 5 – Fixed-wing aircraft service, 1973 section only; 6 – Oilfield maintenance base and helicopter service, 1973 section only.

dah Airport are used for only one month of the year. Pressure on facilities
may produce hazardous situations or environmental deterioration. Only 2200
people per day can climb Ayers Rock. Any more and accidents would be
inevitable. The access roads and viewing platform have so altered local run-
off from the Rock that trees and shrubs have died and soil erosion has been
accelerated (Ovington *et al.* 1973). River boat trippers on the Colorado
River (some 10 000 per year in 1980 compared with 200 in 1960) have so
fouled the sandbar landing and camping sites, which are also eroding under
the new controlled river regime, that severe controls on use have been
introduced (Dolan, Howard, and Gallenson 1974).
   The history of tourist developments and environmental impact can be
modelled as an illustration of the role of accessibility and knowledge on

249

resource use (Fig. 15.5). Inevitably, it would seem, controls upon use have to be imposed and access to the resource rationed. The controls are inevitable not only because of deliberate vandalism, but because of the accidental effects of human use on the area and the difficulties of reconciling a variety of goals in such tourist sites with practical management. Thus the multiple goals of the Ayers Rock–Mount Olga National Park (Table 15.3) could not possibly be all equally achieved. The managers must decide which has to be sacrificed to enable the other aims to be met. So far the solutions have been to restrict numbers by limiting accommodation to 5000.

The pressure upon tourist facilities in the arid lands is increasing. In the early 1970s tourism had increased annually by 25 per cent in Egypt, 11 per cent in Israel and by 10 per cent in Australia. The question is whether the tourist will destroy the resources in the same way as the miners or whether, as the Scottish tourist officer hoped, the scenery can be 'sold' but still retained intact.

Problems facing urban development

Given sufficient incentives, the physical problems of establishing urban settlements in the arid lands can be met by imports of capital, technology, and labour. For the remote mining camps, water can be piped or trucked in. Prefabricated housing or mobile homes can be provided with air-conditioning to create 'refuge' environments for recreation and residential use. But to cope with the social stresses of remote and harsh working environments other strategies have had to be developed.

For remote industrial or mining centres one strategy has been extended shift work alternating with regular free time in the main settled areas. Thus the Israeli potash plant in the barren valley of the Wadi Arabah south of the Dead Sea is worked by a labour force resident at Dimona, a new town associated with Israel's atomic energy research station, 43 km away. The 600 workers at the potash plant have 3–4 weeks shift work then 5 days free. Similar systems operate on the natural gas fields of Moomba in South Australia's Simpson Desert, with intensive shift work followed by free time back in Adelaide 900 km away.

A more complex version of this, called satellization, has been used in the Iranian oilfields to reduce the cost of service provision by removing the many local centres and concentrating their services in one urban centre. Air transport and shift work is used to move staff between their work-places and the centre (Fig. 15.6). The results were a reduction in wages for the company, as employment was reduced from 17 300 in 1957, before the system was introduced, to 9100 in 1969. The companies no longer had to provide basic food supplies to remote centres but left this to private firms, and the cost of maintaining health and welfare facilities was significantly reduced. Economies of scale in the main centre also meant that workers had better access to higher-quality goods and services than before.

Just how far such a system could be applied to other resource uses in the arid lands is difficult to estimate. The monopoly oil industry took advantage of advanced telecommunications to set up remote control systems for oil-well production sites, as well as the helicopters and off-road vehicles for local access. Similar systems have been partly developed in the ranching operations in America and Australia, with workers based on the central station headquarters moving by land vehicle or aircraft to do routine maintenance on remote watering points or survey the condition of livestock and fencing. To some extent the principle also is implicit in the share-cropping and suitcase-farming systems of the commercial grain-farming areas. The principle – that residence may be some considerable distance from work-place – is not unique to the arid lands but it seems most adapted to the low productivity of much of arid space.

The effects of such satellization on the smaller urban centres in the arid lands has been dramatic. From the Saharan oases to the planned country towns of the nineteenth century semi-arid farming frontiers, a combination of lack of local employment opportunities for surplus population, improved transport links to, and the increasing attractions of life styles in, the larger

Plate 15.6 Commercial traveller's guide, Kansas, USA, photographed 1978.
The local 4H youth community welfare organization has listed the names and locations of all local farmers using the cardinal directions and mileages along the grid-iron rural road system. Thus 'Terry Boy 6 Mi N. 4½ W' and 'Howard Plunkett 13 Mi. W¼ S' and, just in case you are lost, 'This is SE¼ SEC. 36-21-41', i.e. Southeast quarter of Section 36 of Township 21 Range 41. Site is 21 miles (34 km) south of Tribune.

251

centres have drawn-off the population. In addition, mechanization of grain-farming systems, grain prices rising slower than costs of manufactured goods, and demands for improved living standards from the rural population have led to increases in farm size, an associated fall-off in rural populations, and decline of local retail trade and service centres (Pl. 15.6). The ultimate stage in the run-down of such centres is captured in Mather's (1972: 247) description of the 'town' of Bill, Wyoming, with a population of two, literally one man and his dog. The man:

> is mayor, fireman, policeman, and postmaster. Forty people from a radius of thirty-five miles [60 km] call for mail at Bill. Based on volume, however, the main urban functions are the retailing of gasoline and beer. Munkres says that 'gasoline gets you across the wide open spaces and beer makes you forget them'.

The urban centres which dot the arid lands show the whole spectrum of characteristics from burgeoning new mining camps to derelict ancient mining sites, from some of the oldest and still thriving cities to some of the youngest and declining townships. That spectrum reflects a sequence of changing appraisals of the arid lands as different societies sought different resources in the harsh environment.

PART IV

Resource management:
some problems and questions

Resource management in the arid lands has been influenced not only by the potentials of the arid ecosystem and skills of the peoples of the arid lands, but also by the cultural and institutional frameworks within which the resource management has taken place. Those frameworks have posed problems for the resource managers which these final chapters aim to identify. The problems concern: first, the ownership of the resources as affected by the systems of land tenure; second, constraints upon the use of vital resources, especially water; third, the administration of the vast and usually sparsely populated arid lands; fourth, the role of human conflicts in affecting arid land development; and fifth, the role of human environmental perception as affecting the attitudes to and management of the arid lands.

Since at least the first three problems will be studied in part through the evidence provided by the legal framework for resource use, some definitions of the relevant legislation are needed. Legislation is the codification of certain rules and standards required by a society. The codification usually implies precise definitions of terms, such as the resources considered, the rules regarding use of those resources, the procedures for settling disputes over that use, and the penalties for misconduct.

From the time of its enactment to the time it may be repealed the legislation reflects a society's codified attitudes – what Toennies (1957) called the *Gesellschaft* system. Separate from, and possibly contrasting with, this was what Toennies identified as the *Gemeinschaft* system – those traditional attitudes and actions inherent in the society. In one sense the *Gesellschaft* system was the codified form of the *Gemeinschaft* system. However, the experience of attempting to apply a fixed codified legal system to the changing real world situations usually created stresses which produced differences between the systems. Over time the gap between intent and achievement widened, until new legislation attempted once more to bring practice and legal purpose together again. In this historical sequence, repeated many times over in the arid lands, we have evidence in the legislation of the intent of resource management and in the story of their enforcement and/or modification over time, evidence of changing societal attitudes to that management and a comment on its relative success. An examination of the sequence of legislation on resource management will illustrate that process by which societies have or have not come to terms with their arid lands.

CHAPTER 16

Resource ownership: the land tenure question

Land tenure usually is defined as the bundle of rights to land which are held by individuals, groups, or institutions in a society. Of concern here will be those parts of that bundle which relate to the ownership of land and how that ownership might affect use of the land. Since many of the current problems of resource management in arid lands relate to the contrasts between traditional systems developed in the arid lands and European systems developed in more humid environments, it is useful to identify the main characteristics of each.

Traditional tenures

Traditional tenures are those which existed before European contact with the area and which may be still in use, possibly alongside later European-derived systems. Despite the great number of such traditional systems, several common characteristics can be identified – characteristics which contrast with European-derived systems and which create some of the present problems of resource management.

Most of the traditional tenures would be designated as *Gemeinschaft* systems, since they were not always codified in written form, more usually being retained within the society by oral traditions, with some members of the groups entrusted with the memorizing of the laws. Such an oral tradition caused problems when brought into contact with the *Gesellschaft* literary traditions of the European systems, particularly in establishing title to land. There are, however, other sources of potential conflicts and difficulties for any effective system of resource management.

Communal ownership

Most traditional land-ownership systems have a significant and often domi-

255

nant component of communal land ownership. With the Australian Aborigines, the indigenes of North and South America, and the nomadic tribes of Africa and Asia, the whole of the land basically belongs to the tribe. No part of that land belongs solely to any one individual. Further, the land *is* the tribe and the tribe *is* the land – they are inherently inseparable. For any one member of the tribe to dispose of his or her share of that land to anyone outside the tribe would not only be impossible under traditional law but would be considered to be an attack upon the very existence of the tribe, since not only might it reduce the resource base, but also remove some inherent portion of the tribe's spiritual being.

Over time these communal rights have been encroached upon and some forms of individual rights have been recognized in some of the arid lands. In Muslim law (*Chari'a*) for example, the community (*umma*) has rights to

all land that has not been improved by individual effort, and especially the arid or rocky lands, which are known as the 'dead lands'.

(Berque, 1959: 487)

Where individuals, by their own efforts had brought parts of such 'dead lands' into production, they could have title. In central Asia before the imposition of Soviet control this was *Amliak* land (Pierce 1960); in North Africa, *mulk* land (Fowler 1972). However, in theory the people could not dispose of such land outside their immediate family. There seem to have been no equivalent modifications in the Australian, American, or South African hunting and gathering nomad groups and the tribe retained all rights. Indeed, where current government polices are trying to encourage retention of traditional systems such tribal groups appear to be firmly rejecting any concept of individual land-ownership rights. As noted in Chapter 10, one of the first 'treaties' to be signed between Australian Aborigines and the governments of Australia, authorized in 1981, gave the Pitjinjatjara people as a tribe the *inalienable* title to 103 000 km² of arid northwestern South Australia. The tribe now has freehold title, but cannot pass on that title to anyone else nor presumably use the land as collateral for loans since no lender could claim the land for unpaid debts.

Religious lands

A further component of the traditional land system is that portion which has specific religious significance and for which special rules apply. Thesiger (1960) noted certain sacred groves in Arabia which could not be used for grazing or firewood. In Muslim law, land may be granted by individuals to the church or some charitable institution as a means of providing institutional income. Such *waqf* or *habous* land is usually granted in perpetuity and is a feature of most of the North African and southwest Asian arid nations.

Part of the Aboriginal land rights campaign in Australia from the 1960s onwards has been aimed at not merely attempting to obtain freehold title of the Aboriginal reserves, which up to then had been administered by the various state government departments of Aboriginal affairs supposedly for the interests of the Aborigines, but also to create as reserves (or at least prevent any changes in use of) areas designated as 'sacred sites'. In most cases these sites are associated with specific topographic features such as boulders, hills, valleys, waterholes, etc. and might lie outside traditional Aboriginal reserves.

Absentee owners

Ownership of the land in the traditional system required neither continuous use nor permanent residence. Ownership rights of nomadic grazing could be exercised even though the owners and their livestock were hundreds of kilometres distant. The lack of any apparent signs of habitation or property markers did not mean that the land was vacant. This fact led to much conflict with invading European settlers occupying what appeared to be empty lands. The conflict was heightened after the mid-nineteenth century when the introduced European systems required residence on, and permanent use of, the lands as a condition of ownership. Even here absentee owners existed, but in the enquiries into the failure of settlement schemes their absence was seen as a major factor in the neglect and assumed mismanagement of the environment.

European tenures

From AD 1500 onwards and particularly during the nineteenth century the expansion of the Western European colonial empires around the world imposed a series of land tenure systems developed in the cool humid lands upon a variety of different ecosystems and cultures, among them the arid lands. Although varying in detail with the nationality of the colonial power, again some basic common attributes relevant to current arid lands management problems can be identified.

Individual ownership

The nineteenth century expansion of the European empires and the United States into the arid lands of the Americas, Africa, Asia, and Australia took place at a time when individual ownership of land was the dominant concept in land tenure. It is arguable, indeed, that the Agricultural and Industrial Revolutions of the eighteenth and nineteenth centuries in Europe and North America, through the means they afforded individuals not only to produce more, but also to control the production of larger areas through mechanization of labour inputs, created situations in which communal labour was no longer necessary and more land could be worked and therefore owned by individuals.

257

Reclamation of the 'commons' or waste lands

At the risk of overgeneralization, it may be suggested that all colonial lands experienced a basic sequence of land tenures. Initially the whole of the territory was claimed by the colonial power and tenure vested in the Crown or equivalent central authority. Next, title was allocated to individuals or institutions as free or low-cost land grants usually for services rendered. Then, the remaining unallocated lands, except for those where indigenous title still was recognized or had been guaranteed by treaty, were offered for eventual individual ownership by sale or as free land provided certain development requirements were carried out. Until ownership had passed into private hands the unallocated areas usually were regarded as waste lands awaiting the realization of their resource potentials by 'development'.

Such a sequence seems to have been true for the arid lands of the Americas, southern Africa, and Australia. In central Asia following the Russian Revolution in 1917 and expansion in the 1920s, original tenures were replaced by state control, reallocation of land rights to actual users, and the imposition of a centrally planned economy in which resource use was, in theory if not always fact, more closely supervised than before. In North Africa, the European colonial powers seem to have given more recognition to traditional occupiers' rights, probably because there was not the pressure for European settlement there compared with other arid areas and the indigenous population was larger and more strongly entrenched on its lands. Only in French Algeria and Tunisia, Italian Libya and Spanish Morocco were significant areas of arid lands set aside for European settlers (Table 16.1). One effect of independence for such colonial territories in the 1950s and 1960s was usually the nationalization of the land and initiation of attempts to remove European settlers or their descendants from land-ownership.

Table 16.1 Tunisian land tenure *c.* 1957

Tenure		Percentage of land
Tribal common lands		41.3
Habous (religious lands)		17.8
State domain		10.6
Mulk (private lands)		30.2
		100%
	Total area	7.3 million ha

Source: Duwaji 1968.

Conflicting tenures?

The existence of two different land tenure systems relating to a common area creates obvious problems of conflicting rights to resources. Whereas one system stresses indivisible communal ownership, the other encourages individual ownership even at the expense of community interests. Whereas in one system the land is a sacred entity integral to the very existence of the society, in the other land is a commodity which can be bought and sold for individual gain. Whereas in one system the link between the owners and the land had an ecological base, that is, the produce of the tribal area was the main sustenance of the tribe, in the other the land need have no ecological significance for the owners, being merely one of a series of potential sources of monetary income.

Small wonder, then, that the last 100 years of resource management in the arid lands has illustrated the clashes of these two tenure systems. The full story of these clashes and the light they throw upon different resource-management systems has yet to be written, but a case study of the evolution of land tenures in the arid lands of Australia and the United States will illustrate some of the conflicts and their resolutions.

Changing land tenures and resource ownership: a case study of the Australian and United States arid lands

The historical context

Over the latter half of the nineteenth century and first three decades of the twentieth century Australian and Anglo-American settlers were attempting to take up arid areas of their two nations under a system of land tenures developed in more humid environments and which was modified in the light of the practical problems of such settlement. Implicit in the original systems applied to both areas was the desire expressed in the respective legislation to create a regular pattern of small family-managed production units in individual freehold* ownership. By 1970 only about 5 per cent of the Australian arid lands and 41 per cent of the United States arid lands had been so allocated (Table 16.2 and Fig. 16.1). The largest area of both arid lands was in leasehold† tenure while reserves§ occupied between 7 per cent and 13 per cent and in Australia almost a fifth of the area was still unused in Crown ownership.

* Freehold title gives the owner full and free rights to the bulk of the land resources (except possibly the minerals) in perpetuity after the initial payment for, or grant of, title to the land. [Australian term is alienated, in USA patented, land.]
† Leasehold title gives the lessee limited rights for a specified time to specific land resources in return for payment of an annual fee (rent).
§ Reserves are lands owned by the government and preserved for public use or special purposes and protected from exploitation by individuals for varying periods of time.

259

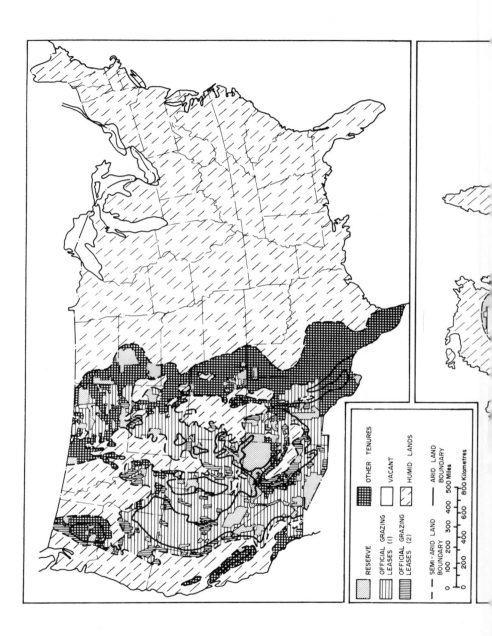

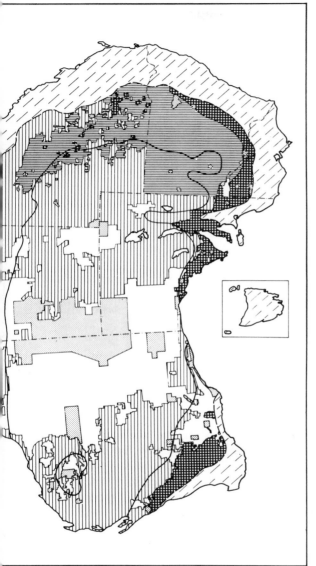

Fig. 16.1 Land tenure in the arid lands of Australia and USA *c.* 1970. *Sources:* Heathcote 1969, USA 1970.
Key: Official grazing leases (1) in USA includes leases on Bureau of Land Management lands and state lands with some possibly on National Forest lands. In Australia they are leases the size of which is not restricted to 'home maintenance' or 'living' areas (i.e. not tied to single family production units)
Official grazing leases (2) in USA are leases within the National Forests. In Australia they are leases the size of which is restricted to 'home maintenance' or 'living' areas (New South Wales and Queensland only). Reserves include Indian lands, military and recreational lands in the USA and Aboriginal reserves, scientific and recreational lands in Australia
Other tenures are mainly patented lands in the USA and alienated lands in Australia

Table 16.2 Land tenure in arid Australia and the USA *c.* 1970

	Percentage of the arid lands[a]	
	Australia	*USA*
Freehold	4.9	40.7
Leased	70.2	46.2
Reserved	6.8	13.1
Vacant	18.1	0
	100.0	100.0
Arid lands (as % nation)	81.0[a]	34.0[a]
Other estimates (arid as % nation)	75.0[b]	30.0[b]
	64.0[c]	30.0[c]

Sources:
[a] Fig. 16.1 (Meigs 1952, *extreme arid, arid*, and *semi-arid* areas).
[b] Paylore and Greenwell 1979 (*arid* and *semi-arid*).
[c] Rogers 1981 (based on UNESCO 1977 map *World Distribution of Arid Regions* 1: 25 000 000, including *hyper-arid, arid*, and *semi-arid*, but not *sub-humid*).

This gulf between the goals and the achievements of resource-management and land-settlement policies reflects the original misconceptions about the productivity of the arid lands. Experience of the problems of arid lands management and changing philosophies of general resource management led to the modification of policies and the introduction of modified tenure systems such as the leaseholds and reserves. The details of such modifications require explanation.

Coping with low unit area productivity

A fundamental problem of the initial tenures was that the unit size of land-ownership proved inadequate to support the family for which it was intended. In the USA the prior Spanish land grants were initially of large areas (often from 6000 to 400 000 ha) and the problem did not apply, but the 1862 Homestead Act which was intended to apply to all the unallocated lands (including the arid areas) set a limit of 65 ha (160 acres)* and this subsequently had to be enlarged. In 1902 an enlarged homestead of 130 ha (320 acres) was allowed in the arid western states. In 1904 on the grassed sand dunes of Nebraska the Kinkaid Act allowed 259 ha (640 acres) homesteads, while in 1916 the Stock Raising Homestead Act made 259 ha homesteads

---

* The imperial measurements of the original legislation are indicated in parentheses.

available for the western states. In Australia the picture was complicated by variations of detail between the five mainland states, but essentially the policies were the same – initially 'more and smaller is better' (Williams 1975). Here also sizes were increased over time, in South Australia from 32 ha (80 acres) to 65 ha, later 130 ha and even up to 1619 ha (4000 acres) in the 1920s to provide what were considered to be economic units.

An alternative strategy by which productivity might be improved was the encouragement of irrigation. The expansion of irrigation in the arid United States already has been noted in Chapter 13 (p. 203). That expansion was based in part on official encouragement through special land tenures as well as government engineering works. Thus in 1877 the Desert Land Act allowed a homestead of 259 ha to be selected, provided a portion of it could be irrigated. Eventual title depended upon the development of irrigation, although loopholes in the laws allowed speculation without development. The conditions were tightened and the size of land was reduced to 128 ha in 1890. The intention, however, remained to encourage bona fide irrigation. In Australia several officially sponsored irrigation schemes were located along the Murray River and its tributaries and the bulk of the 10 per cent of the irrigated area within the arid lands is contained within state government planned developments (Table 16.3). While it is doubtful whether major government funds will be allocated for any further irrigation developments in either country, there is no doubt of the original intention.

For a brief period in the late nineteenth century governments in both countries recognized that tree planting might ameliorate the arid climate and made provision accordingly in their land tenure systems. Tree planting was thought to increase rainfall as well as provide shade and protection from drying winds. When the future timber potential was added the prospect had a particular attraction for the semi-arid lands. In the United States the 1873 Timber Culture Act gave settlers 65 ha free provided 16 ha were planted with trees, and from 1899 onwards portions of the arid lands were retained in federal ownership as National Forests, although this was more a protective measure of existing timber rather than a tree-planting programme. In South Australia tree plantations were created in 380 mm rainfall country in the 1870s for both future timber and climatic modification purposes. In this latter context the Saharan reclamation schemes (Calder 1951) and work of the Men of the Trees Society is an international example of more recent attempts to modify local climates by tree plantings.

A final strategy to cope with low productivity was to allow piecemeal use of *some* of the land's resources at a fraction of the cost for access to the whole suite of resources. In both countries this was applied particularly to grazing land tenures, but the concept is implicit in most of the current arid land tenures (Fig. 16.2). In Australia the first European pastoralists in the arid lands were illegal occupants (squatters) unwilling to pay £0.1–£0.25 per ha for land which would at best support only a sheep for say 3–4 months

Table 16.3 Irrigation in Australia

A. Land use on irrigated areas (1975–6)

	Percentage of irrigated area
Pastures (inc. lucerne)	62.4
Cereals	17.1
Sugar-cane	5.0
Fruit, vegetables, vines	11.0
Other crops	4.5
	100.0

Total area irrigated   1 474 900 ha = 10.1% of cropped area.
Irrigated area in arid lands ≈ 10%.

B. Source of water for irrigation *c.* 1970s

	Percentages of water supplied			
	Surface	Underground	Urban (reticulated)	Totals
Government schemes	51.2	—	1.2	52.4
Other schemes	32.9	14.7	—	47.6
Totals	84.1	14.7	1.2	100.0

Source: *Australian Year Book* 1979.

of the year. From the mid-1830s a system of depasturing licences and grazing leases allowed use of only the grazing resources for a minimal rent (£0.005 per ha). Such a system has remained the dominant tenure of the arid lands ever since (Table 16.2). In the USA arid grazing leases seem to have been used in Texas from 1883 onwards, but despite proposals for leases (at rentals of 5 cents per ha per year) from the cattlemen's National Livestock Association in 1900, and provision for seasonal grazing leases in the National Forest reserves from 1907 onwards, pastoralists were expected until 1934 to survive on 259 ha at most. In effect the arid ranges from the 1860s to the 1930s were being grazed illegally and free of any charge. In 1934, almost exactly 100 years after the principle of grazing leases had been established in Australia, the American *de facto* pastoralists were allowed legal title to grazing leases at nominal rents under the Taylor Grazing Act.

The principle of separation of 'slices' of resources for separate allocation shown on Fig. 16.2 is implicit in most non-freehold tenures in the arid lands, from wildlife reserves, military reserves which may allow grazing for part of the time, to mining rights which may underlie pastoral leases. The evolved system provides legal title and rights to use for a portion of the total resources without the cost of acquiring the other relatively unattractive components of the sandwich of resources.

Coping with productivity variations over space and time

In both Australia and the United States there is evidence of some attempts to modify tenure systems to cope with the variable arid climate and associated fluctuating productivity of the land. To allow farmers to seek off-farm

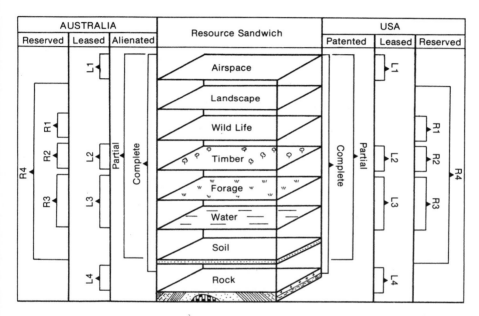

Fig. 16.2   Allocation of the 'Sandwich of Resources' in Australia and USA. *Sources*: Heathcote 1969; USA 1970.
    The various facets of the resources of the environment have been separately allocated in both Australia and the USA
    Key: Australia: Reserved lands: R1 – Fauna reserves; R2 – Forest reserves; R3 – Grazing reserves, e.g. travelling stock routes; R4 – National Parks.
    Leased lands: L1 – Military leases; L2 – Timber leases; L3 – Grazing leases; L4 – Mining leases.
    Alienated lands: Complete – all facets owned. Partial – more usual situation where mineral rights are retained by the Crown for separate allocation
    USA: Reserved lands: as for Australia
    Leased lands: as for Australia
    Patented lands: as for Australia except that the complete range of resources is usually owned, and partial ownership with minerals owned by the State or federal authorities separately would be the exception.

Table 16.4 Official drought relief for the arid lands in Australia and USA

A. Australia – drought relief 1965–7

		$Amillion
Commonwealth	– loans to states at low interest	108.8
	– grants to states	60.0
	Total	168.8
	Arid lands share (approx. one-third) =	56.2

B. USA – drought relief 1954–6

	$US million
(i) Federal payments to semi-arid states:	
Colorado	27.6
Kansas	30.2
Montana	5.7
Nebraska	1.7
North Dakota	12.8
Oklahoma	44.9
South Dakota	6.5
Texas	181.7
Wyoming	5.6
Total semi-arid states	316.2
(ii) Federal payments to arid states:	
Arizona	2.8
Nevada	3.3
New Mexico	28.6
Utah	4.6
Total arid states	39.3
Grand total arid lands	355.5
Grand total as % of total federal relief	90.5%

Sources: Australia – Heathcote 1967; USA – Warrick 1975.

employment during periods of drought or other economic stress the residence requirements of homesteads in both countries were often relaxed. In some cases permanent residence could be off-farm provided it was within a working distance. The droving of livestock to seasonal semi-arid pastures was allowed for in Australia by provision of official travelling stock routes up to 60 m wide (similar to the British drove roads and Spanish *mesta* trails – the latter being 90 varas/metres wide). In droughts these tracks in theory became escape routes for starving livestock. The Taylor Grazing Act in USA similarly recognized prior rights of seasonal stock movements from the winter ranges of the arid basins to the summer ranges of the National Forests and made allowance for such movement on grazing leases.

In both countries the state and national governments have provided massive drought relief to the arid lands over the years (Table 16.4). Most relief has been given to encourage the resource managers to stay on their properties in the hope that conditions will improve. Relief has been as remission of rents of leases, of loan repayments, as low-interest loans for seed or livestock replacements, or even as emergency transport for the evacuation of livestock. In a sense this relief has been a regional subsidy from the national treasury as noted in Chapter 12. The question remains whether the nation benefits or loses by such subsidies.

Conclusion

The last century has seen the modification of the humid-based land tenure systems introduced into the arid lands of Australia and the USA. For over half of the area in both nations, the land is still owned by, and ultimate responsibility for resource management still rests in, the government. As we have seen, this management is exercised through the various leasehold systems which form the largest single category of tenures in the two arid lands. In addition, however, government controls are exercised through the reserved lands and to explain the areas currently under these latter tenures we need to consider the changing role of government in general resource management.

CHAPTER 17

Resource-management systems

Management of the arid lands' resources requires careful allocation of the scarcest item – water, strategies to cope with the vast and relatively empty spaces in the arid lands, and the ability to plan for multiple resource management made possible by the differentiated tenure systems noted in Chapter 16.

Allocating the scarcest resource – water

Traditional systems

The history of water allocation in the arid lands offers an impressive array of ancient legislation on the classification of the water resources and priorities for its use:

> Assyrian irrigation laws distinguished between rain-water and water from natural springs, wells or cisterns. Regulations dealt with the rights and duties of landowners whose properties were irrigated from a common source; they provided for cooperation in maintaining supply, keeping canals free from silt and pollution, and ensuring that those farthest from the source received a fair share.   (Drower 1956: 532)

The Code of Hammurabi (1800 BC) set the priorities for use of water in Babylonia as: first, drinking water for men and livestock; second, domestic uses; third, irrigation; and fourth, navigation. Compensation for misuse of the resource was quite specific:

> If a man open his canal for irrigation and neglect it and the water carry away an adjacent field, he shall measure out grain in proportion to the size of the adjacent field.   (Vallentine, 1967: 26)

268

Within the irrigated areas of Asia and North Africa, the complex frameworks for water allocation have been maintained and elaborated over time. Reviewing Muslim water laws in 1973, Caponera (1973: 28) noted that:

> Customs governing water ownership in arid zones are dominated by the fact that in desert areas water constitutes the main object of real property. As water becomes scarcer the land proportionally becomes an accessory to it, contrarily to what is the case in European legislation.

While the traditional 'right of thirst' – the right of access to drinking water for human and animal needs – still applied for all communal sources, in the Shi'ite areas (some 9% of the Muslim faith) this right did not apply to private water sources. Generally, the larger the source of supply the fewer the restrictions upon use. Thus lakes were for communal uses whereas small streams might be solely in private ownership. Private rights usually accrued as a result of past development activity, such as well-digging, stream diversion, and canal construction. Allocation of the water was carefully supervised – usually by a specific official (*amin* or *mirab*) – and might be by volume or period of flow, with subdivisions of time down to half an hour quite common within a 14-day rotation period. Water from *qanats* was owned communally and allocated on a roster system by similar time periods. While most of the Muslim nations since 1968 have set up centralized bureaucracies to administer national water policies, at village level traditional systems of allocation seem relatively unaffected by such innovations.

Riparian versus prior appropriation systems

Contemporary water law in those arid lands settled by Europeans appears to reflect components of two separate systems of water rights. Although both systems appear to have been derived originally from the Mediterranean region, the different routes to the arid lands reinforced inherent variations so that now they offer different potentials for resource use (Table 17.1).

For the allocation of water in the arid lands the prior appropriation system appears most suited. In this system water can be diverted and allocated for irrigation purposes from one river basin to another. It is a saleable commodity separate from the land. It can be completely consumed and the system encourages complete use of the resource, since incomplete use results in forfeiture of rights to the unused amounts. By comparison, the riparian system seems best adapted to humid lands where water is neither a scarce nor excessively localized resource.

Although there is no global survey available of the different characteristics of the operative water law, Hamming's map of the pattern of the two systems in the USA (Fig. 17.1), and the Canadian and Australian experience, point to a somewhat similar sequence by which systems of water management based upon prior appropriation or drastically modified riparian

Table 17.1 Riparian and prior appropriation water rights

Components	Riparian rights[a]	Prior appropriation rights[b]
Water ownership	Inherent in the land	Only from prior use or statutory rights
Location of land to which water rights accrue	Must adjoin stream	Need not adjoin, i.e. can be distant from stream
Use of water	Diversion for 'natural uses' only; non-consumptive	Consumptive; prior rights may claim whole
Amount of water used	No limit in theory	Fixed – by prior right – by licence
Duration of water right	Infinite; does not lapse if not used	Finite; lapses if not used

[a] Derived from British Common Law.
[b] Derived from Roman or Canon Law.
Source: Hamming 1958.

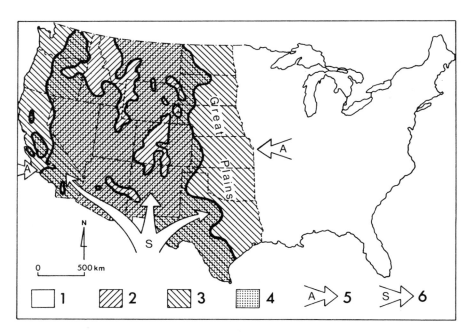

Fig. 17.1 Water rights in the USA. *Source*: Hamming 1958
The dominant water rights for the states in the arid USA are indicated.
Note that state borders in the eastern humid USA are not shown.
Key: 1 – Riparian rights apply; 2 – Prior appropriation rights apply; 3 –
A mix of riparian and prior appropriation rights apply; 4 – The arid lands;
5 – Anglo-American settlement approach routes; 6 – Spanish settlement
approach routes

rights have come to be adopted in the arid lands. That is not to say that disputes over water have ended. As we saw in Chapter 13, the allocation of Colorado River water in the western USA has been long disputed. In Australia, while the allocation of flows of the Murray River has been decided between the three adjacent states (New South Wales, Victoria, and South Australia), control of the *quality* of the water is still in dispute. However, the legal context of the disputes has come to be that of prior appropriation with specific amounts and qualities of water being allocated among competing users.

Administering the arid lands

Given the vast size of the arid lands, their low population densities, and their thin scatter of urban centres, governments face not only security problems but also the problem of providing some kind of representative government.

Securing the lands

The security of the arid lands has been effected traditionally from fixed fortified bases centred on the oases with occasional forays into the 'empty' lands by mobile self-supporting patrols. The Assyrian Emperor Ashurbanipal (669–626 BC), campaigning against the bedouin near Damascus, controlled the oases and thus, in theory, controlled the intervening space:

> in every place where there were springs or wells of water, I set guards over them, depriving them of the water to keep them alive. I made drink costly to their mouths; through thirst and deprivation they perished.
> (Drower, 1956: 527)

However, as T. E. Lawrence demonstrated in the same area during the First World War and the British Army's Long Range Desert Group showed in North Africa in the Second World War, a highly mobile self-sufficient group could strike effectively against such fixed bases and retreat safely into the empty spaces. In northeast Brazil's arid scrub (*caatinga*), guerrillas in the 1890s held out for ten months against superior government forces (Cunha 1957). In theory air surveillance should make such exploits less likely nowadays, but the contemporary (1981) conflict in Afghanistan illustrates the difficulties of controlling guerrillas in arid mountains from cities on the plains even *with* air power.

Until the creation of the newly independent nations of North Africa from the 1950s and 1960s onwards, the demarcation of political boundaries in the arid lands varied from heavily fortified entrenchments, such as the Israeli-Egyptian border of the 1950s, to the invisible cartographer's line on the shifting sands, such as the boundary between French Algeria and the Spanish Sahara, which had been fixed in 1904 as the meridian of 8° 40' and for a strip

271

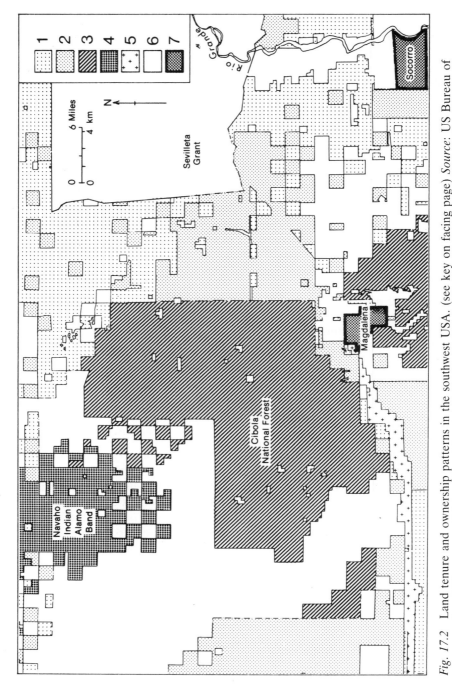

Fig. 17.2 Land tenure and ownership patterns in the southwest USA. (see key on facing page) *Source:* US Bureau of Land Management Maps, 'Magdalena' and 'Socorro' sheets, scale half-inch to one mile, dated March–April 1969, and USA 1970.

of 42 km was marked by only two pillars! The creation of the new nations, however, focused attention on territorial boundaries. Despite the illogicality of many of the colonial divisions, these boundaries have become more obvious on the ground and more effective barriers to traditional movements.

Governing the lands

The low population densities of the arid lands often provide an inadequate tax base for their own government. As a result administration of such areas is subsidized from the national treasury, as we have seen in Chapter 12 (pp. 184–5).

Over large areas of the arid lands in Australia and the western USA there is no effective local government as the size of the population is inadequate. Administration is carried out by several often unrelated government departments or agencies. In 1953 a map of Australian local government areas showed both the arbitrary nature of boundaries drawn by meridians and latitude across empty space and that portion of western New South Wales and northern South Australia which had *no* local government. In these cases, 33 per cent of New South Wales and 80 per cent of South Australia were governed directly from the state capitals by the various departments of roads, railways, agriculture, health, education, and in New South Wales by the Western Lands Commission for the pastoral leases. In the western USA the chequerboard land survey and tenure system means that any region is allocated among several private and official management agencies (Fig. 17.2). Regional planning in such a context must be a nightmare.

The official role in arid resource management

Even a cursory glance at the present composition of land tenures in those parts of the global arid lands which were subjected to European settlement from the nineteenth century onwards (see Table 16.2) reveals that central governments directly or indirectly still own large areas. In addition to this responsibility for management based upon ownership, central governments

Key: 1 – Public domain, unallocated; 2 – State lands; 3 – Federal National Forest lands; 4 – Indian lands or reservations; 5 – Federal Bureau of Land Management lands; 6 – Patented lands; 7 – Urban areas
Notes: The area lies within New Mexico and includes a portion of a Spanish (Sevilleta) land grant. Except for cultivation along the Rio Grande, the bulk of the map area was grazed by cattle and sheep under various forms of leasehold tenures from the various owners (private and government agencies)
In 1968 32 per cent of New Mexico was still federal land. This represented 10.2 million ha, of which the Department of the Interior administered 5.3 million ha (mainly through the Bureau of Land Management), the Department of Agriculture administered 3.6 million ha and the Department of Defense (mainly through the Army) administered 1 million ha

have taken over an increasing responsibility for resource management even over arid lands in private ownership. Both this continued governmental ownership and extended powers of resource management need explanation.

As we have seen in Chapter 16, when allocation of the whole array ('sandwich') of site resources failed, segments of those resources were allocated separately (by leases or licences) with ultimate ownership of such segments as well as the remainder being retained by the government. Through the creation of a series of government agencies or expansion of departmental responsibilities, separate portions of the arid lands resources were managed.

In addition, the occupation of some of the arid lands coincided with the growth of conservation ideologies in which it was argued that certain arid land resources should be retained in public ownership for the benefit of the nation rather than individuals. Such resources were identified as the archaeological sites of prior civilizations, to be preserved on historic, cultural, and educational grounds; lands still inhabited by indigenous groups whose land title was to be guaranteed on cultural grounds as reserves for their use alone; areas of natural beauty, particularly the more spectacular arid landforms, to be retained on aesthetic and educational grounds; areas of scientific interest where rare plants or animals should be retained, also on educational grounds; and finally, some recreational areas where the terrain or some aspect of the resources offered the general public recreational facilities. Separate from the above were the specific needs of the military forces for space to test equipment and troops, or carry out secret or dangerous activities. In this case direct ownership of empty and remote parts of the arid lands provided that space.

The net effect of such pressures to retain land in government ownership has led to a significant proportion and variety of resources in the arid lands being retained under official management. Within the arid areas of the USA official controls may apply to as little as 2 per cent and to as much as 86 per cent of the area (Table 17.2). Within such areas a multiplicity of agencies exercises control over a variety of resources as we have seen above.

This direct official involvement in resource management has been complemented from the 1930s onwards by increasing official intervention in private resource use. Again this has been justified by a conservation ideology where private misuse of resources was seen as leading to national losses. As noted in the previous discussion of rain-fed agriculture and ranching, the serious impacts of both upon soil erosion rates led to official intervention in Australia, South Africa, and the United States, usually merely as rehabilitation but occasionally requiring changes in the intensity of land use.

The 'policing' of private resource management is often more apparent than actual on the ground. In the Australian and United States' experiences, it is rare that offending managers receive the full penalties of the law since it has been relatively difficult, prior to satellite imagery being available, to prove significant deterioration attributable to specific management systems. The new imagery will assist the monitoring process, but whether governments will be anxious to punish offenders remains to be seen.

Table 17.2 Official resource management responsibilities in arid western USA c. 1968

State (area million ha)	Federal land as % state	Federal Land Administration (% state by agencies)										
		Dept. of Interior					Dept. of Agriculture			Dept. of Defense		
		1	2	3	4	5	6	7	8	9	10	11
Arizona (29.4)	44.6	17.6	2.1	2.2	1.9	0.1	15.7	.	.	3.5	1.4	.
New Mexico (31.5)	33.9	17.1	0.1	0.3	0.3	0.5	11.8	—	.	0.1	3.4	0.1
Nevada (28.4)	86.4	68.4	3.1	0.4	1.7	.	7.2	—	—	4.2	.	1.6
Texas (68.1)	1.8	—	0.1	0.5	.	—	0.5	.	.	.	0.2	0.4

Notes: 1. Bureau of Land Management; 2. Fish and Wildlife; 3. National Park Service; 4. Bureau of Reclamation; 5. Bureau of Indian Affairs; 6. Forest Service; 7. Soil Conservation Service; 8. Other; 9. Air Force; 10. Army; 11. Other (includes Atomic Energy Commission, Navy, Corps of Engineers); Dot indicates area too small to be included; — indicates no land in this category.
Source: USA 1970, Appendix F.

Conflicts over resources: the conflict between the desert and the sown

In 1953 Spate noted three 'perennial motifs with a strong geographical backing which run through the course of human history'. They included the contrasts and often conflicts between townsmen and farmers, and seapower versus land power. But

> most striking, perhaps, is the secular struggle of peasant and nomad on the frontiers of the desert and the sown, which have shifted back and forth with climatic changes, with the rise and decay of strongly organized states, with changes in the technology of war and peace. . . . So spectacular have been the changing fortunes of this struggle that some [Ibn Khaldun and A. J. Toynbee] have seen in it a main key to the course of history.　　　　　　　　　　　　　　　(Spate 1953: 16)

Ibn Khaldun and Toynbee saw the conflict mainly as a southwest Asian and North African phenomenon, where the contrasts in apparent wealth and life style between the oases and river-lands and the desert hinterlands were so marked and where cycles of feast alternated with famines from droughts or plagues – whether of insect pests or plant diseases. Others have recognized the same conflict as the stimulus for unrest in Latin America, where it was claimed the conquering Spanish pastoralists ousted indigenous agriculturalists from the fertile valleys in the sixteenth to seventeenth centuries and laid the foundations for the twentieth century

> conflict between the pastoral interests of the large landed proprietors and the pressing agricultural necessities of the lowly rural masses of the population.　　　　　　　　　　　　　(Lynn-Smith 1969: 25)

The 1917 Mexican agrarian revolts, as well as conflicts in Peru in 1950 and in Bolivia in 1953 were identified as examples of this conflict.

The impacts of these past struggles have left their mark upon the arid lands: the destruction from the contemporary conflicts threatens the survival of any human settlement in some of those arid lands. Some of the most devastating conflicts have occurred in the cockpit of the Sinai Peninsula and on the Negev frontier.

The Sinai cockpit

Stand at the meeting ground of Africa and Asia, on the Sinai Peninsula, say on the rocky slopes of the Wadi El Arish, and count the invading armies as they would have passed before you from the beginning of recorded time:

2000–1700 BC　　Amorite nomads from Syria attacking Egypt
1700 BC　　　　　Hyksos chariots and horsemen from central Asia attacking Egypt

1500 BC	Egyptian Army of Thothmes I *en route* to attack Syria and Mesopotamia
1190 BC	Philistines attacking Egypt
670 BC	Assyrians attacking Egypt
525 BC	Persians under Emperor Darius attacking Egypt
332 BC	Greeks under Alexander to and from Egypt
312 BC	Egyptians under Ptolemy on their way to attack Jerusalem
217 BC	Egyptians attacking Syria
100 BC–AD 700	Various Roman armies but generally the *Pax Romana*
AD 1250	Mongols attacking Egypt
1798–1801	French, originally under Napoleon, from Egypt attacking Syria
1915–17	First World War – Allied–Turkish conflict
1956	Israeli forces to Suez Canal
1967	Israeli forces again to Suez Canal and occupation of Sinai
1973	Egyptian counter-attack across Suez Canal
1975	Ceasefire ends hostilities between Egypt and Israel and Israeli withdrawal planned

While the exactness of the early dates can be questioned, there is no doubt that the armies did march in the sequence indicated, and, unlike most modern armies, did not carry extensive rations but lived off the land along the way. Such constant and intensive foraging together with the associated slaughter and destruction must have had massive impacts on the landscape and its inhabitants. The Assyrian Emperor Sennacherib (704–681 BC) gloated over his destruction of Babylon:

> I threw down the city and its houses from the foundations to the summits; I ravaged them and allowed them to be consumed by fire. I knocked down and removed the outer and the inner walls, and the temples and all the brick-built ziggurats and threw the rubble into the Arahtu canal. And after I had destroyed Babylon, smashed its gods, and massacred its population, I tore up its soil and cast it into the Euphrates so that it was carried by the river to the sea.
>
> (Wellard 1973: 147)

Omitted was the more favoured method of destroying the enemy's agriculture, deliberately strewing salt on the fields. In the face of such deliberate and wilful destruction we can only marvel that any settlement survived the bloodied hand of time.

The Negev frontier

Fringing the Sinai Peninsula on the northeast, the Negev has experienced a sequence of land settlement which mirrors many of the political and environmental stresses which have faced and continue to face this portion of the

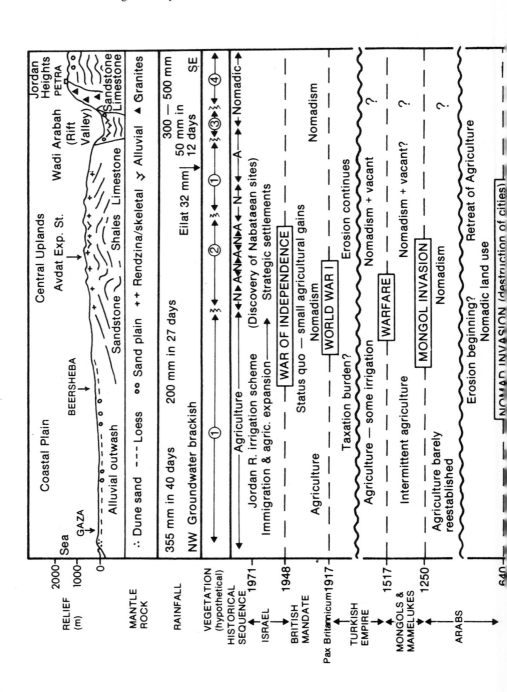

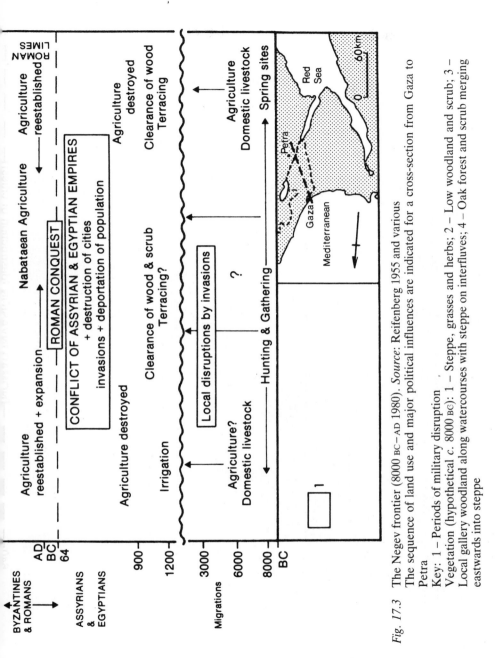

Fig. 17.3 The Negev frontier (8000 BC–AD 1980). *Source:* Reifenberg 1955 and various

The sequence of land use and major political influences are indicated for a cross-section from Gaza to Petra

Key: 1 – Periods of military disruption

Vegetation (hypothetical *c.* 8000 BC): 1 – Steppe, grasses and herbs; 2 – Low woodland and scrub; 3 – Local gallery woodland along watercourses with steppe on interfluves; 4 – Oak forest and scrub merging eastwards into steppe arid lands (Fig. 17.3). The sequence shows the advance and retreat of successive systems of sedentary resource use, many of which in Glueck's telling phrase were 'reduced in time to scrap heaps of artifacts' (Glueck 1959: 97). The transformation of the prehistoric hunting and gathering economy into agriculture at favoured water points was disrupted by migrations and invasions of northern peoples. Then the expansion of irrigation and clearance of the scrub woodland for agriculture was disrupted by Egyptian and Assyrian invasions.

The Roman conquest brought a period of relatively peaceful development behind the line of forts – the Limes – east of the Jordan Valley. This was

Plate 17.1 Moshav Yotveta, Israel. *Source:* Pantomap Israel Ltd.
> View northeast across the settlement to the Wadi Arabah and Jordan in the distance. On the plateau either side of the settlement can be seen the fortifications defending the central housing areas. In the top left is the main irrigated area beyond the north–south road along the Wadi Arabah.

the time when the Nabataean irrigation system expanded into the central uplands of the Negev. The collapse of the Roman Empire brought a period of disruptive campaigning through the area as various empires rose and fell. The period of the Turkish Empire, 1517–1917, brought some revival of agriculture alongside the nomadic pastoralism, but it has been claimed that the burden of taxation prevented extensive agricultural expansion.

The British Mandate of Palestine brought relative peace, new Jewish settlements, and the hopes for a Jewish National State. The 1948 insurrection or war of independence, depending upon which contestant's view is adopted, led to the British withdrawal, declaration of the State of Israel and the expulsion or flight of Palestinian Arabs from Israeli-occupied territory. Further conflict between Israel and Jordan in 1967 resulted in Israeli occupation of Jordanian territory on the west bank of the River Jordan and the flight of some 150 000 inhabitants to become the Palestinian refugees who still occupy United Nations camps in Jordan, Lebanon, and Syria.

Within the State of Israel the period since 1948 has seen feverish activity to re-establish agriculture, not least in the arid Negev, with the aid of the most modern scientific techniques. These have included adoption of ancient Nabataean irrigation methods (Ch. 13, p. 190) and a national water budget (Ch. 5, p. 73), alongside experiments in communal resource management from moshova to kibbutzim in fortified settlements – fortified by irrigation against the desert and by military force against human adversaries (Pl. 17.1). The peace treaty (Camp David Accord) of 1977 with Egypt has allowed freer movement and co-operation on arid land development between the nations, but on the eastern and northern borders the 'desert' is only a rifle-shot away.

CHAPTER 18

Perception of arid resources: contrasts and ambiguities

The variety of resource ownership and management noted in Chapters 16 and 17 was explained in part by the distinctive challenges posed by the arid resources. The aim of this chapter is to demonstrate the role of human perceptions of those resources in further explaining that variety.

Full documentation of the role of environmental perception in arid lands management is beyond the scope of this chapter, but a simple model will allow some points to be made. In 1620, in his book *Novum organum*, the English philosopher Francis Bacon described four idols, or false gods, which led men away from the path to Truth (Fig. 18.1). Each of those four idols illustrates a different influence of perception upon knowledge of, and attitudes towards, the arid lands.

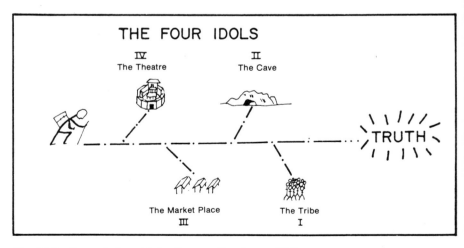

Fig. 18.1 Bacon's four idols. *Source*: Heathcote 1976.
See text for explanation

The Idol of the Theatre: the constraints of language

Bacon suggested that the limitations of language and the problems of semantics affected human knowledge and understanding of the global environment. As we saw in Chapter 2, human understanding of the term 'desert' appears to have changed over time so that study of historical evidence would need to be aware of the past interpretation of the term.

There is also evidence of a complexity of descriptive terms applied to phenomena in arid lands which suggests an enlarged spectrum of human concern for and knowledge of certain aspects of the environment. Thus, Bertram Thomas identified sixteen different terms applied by the nomads to types of sand formations in Arabia in the 1930s – terms which had particular relevance to the nomads in that they signified different qualities of grazing resources and trafficability (Table 18.1). European occupation of the arid lands required the naming of unfamiliar phenomena – of river-beds where water only rarely flowed, of the variety of sand forms, of unfamiliar life forms, of strange systems of resource management. As a result old terms were changed, foreign terms adopted, and new terms coined. 'Creek' was transformed from a tidal inlet to a dry watercourse in Australia; the Arabic *barchan* was widely adopted for the crescent dune form; the Spanish *arroyo* was adopted by Anglo-Americans for the wadis of the southwestern USA; and 'ana-branch' was coined for a distributary which rejoined a river on the semi-arid plains in Australia.

Quite apart from the obvious difficulty in applying a definitive terminology to the components of the arid environment, there was the separate prob-

Table 18.1 Topographical terms for sand formations in southern Arabia

Term	Description
huta	Gorge between sand-hills or sand ridges
markhtun	Wall of sand transverse across a 'huta'
raidfa	Gentle curved slope of 'markhtun'
rakib	Flat roof of sand-hill
rubadh, argab	Drifts of small sand-banks
kharaiyim	Corridors of steppe vegetation between sand-hills
mukhdat	Offshoot of wadi emptying into sands
masila	Soft sandy bed of wadi in steppe
ga'aida	Large sand-dune
sinun	Conical sand-dune
shansuba	Lofty sand dune
barkhan, hugua	Horseshoe sand-dune
urug	Sand-dune ridges
zibara	Hard sand (good going)
ath-ath, beth-beth	Soft sand (bad going)
asfila	Fall of wadi where it is lost on edge of sands

Source: Thomas 1932.

283

lem of terms which were claimed to be not only descriptive but explanatory. The debates over the meanings and relevance of terms such as 'pediment' and 'inselberg' in the arid lands are particular examples (McGinnies, Goldman, and Paylore 1968; Twidale 1981). The confusion arises because of uncertainty as to whether the terms are descriptive merely of the forms or imply also the process which produced the forms. In linguistic studies, such general problems are still unresolved (Whorf 1956).

The Idol of the Market-place: the contemporary philosophies

For Bacon the Idol of the Market-place represented the seductive influences of the contemporary philosophies or the 'climates of opinion', those ways of interpreting the contemporary world which had the majority of followers and which influenced most decision-makers. Such philosophies interpreted the facts in particular ways and set goals for resource management. Two aspects of the role of these philosophies will be examined here: first, the role of contemporary development theory in affecting general attitudes to and management strategies for the arid lands; and second, a case study of the Great Plains of the USA to illustrate how changing philosophies appear to have affected private and official management policies over the last 150 years.

Development theory

The word 'development' is perhaps one of the most common terms in current global political and economic communications. Development appears to be the aim of most governments, from the largest capitalist democracies or centrally planned economies to the smallest dictatorships. The origin of the term seems to be associated with the Darwinian theory of natural selection which was adopted as a form of Social Darwinism – where societies survived according to their inherent ability to respond to both the physical environment and other societies. Those societies which benefited from the Industrial Revolution were able to manipulate the physical environment more intensively, tap more resources, and amass a larger material wealth, thus dominating the less industrialized societies. The key to survival of all societies was seen to lie in the efficient realization of the latent benefits of their resources – in other words, development.

The means by which such development should be achieved have been debated fiercely by advocates of *laissez-faire* capitalism on the one hand, and of state controls by central bureaucracies on the other, while an array of mixes of private and public management have been suggested in between. The contrast between rich and poor nations has been explained as the result of exploitation of the resources of the poor by the rich, whether extraction of cheap raw materials in return for expensive manufactured goods, or use of cheap labour to produce manufactured goods, or by use of the national environments as sinks for pollutant by-products of manufacturing processes no longer allowed in the already polluted richer nations. Partly in an attempt to reduce that contrast and partly to achieve specific political goals, transfers of wealth as 'foreign aid' have been called for and justified. Officially the aim of such transfers has been to further the development of the recipient nation by this transfer of capital, and thereby enable it to exploit more efficiently its own resources whether of land or labour. In addition, however, foreign aid has been used for political purposes to support power blocs and strengthen allies.

While there is no doubt that the philosophy of development is inherent in most official arid land resource-management policies, there is also no doubt that such policies have been criticized seriously. Apart from the obvious criticisms concerning the political motivation and the fact that the inter-governmental transfers, which are the main form of the aid pro-grammes, lead to money either finding its way into private pockets, or being spent in unproductive 'show' projects, there is evidence that the aid remains centralized and urbanized and is not dispersed into the rural areas where it might be most beneficial. Other serious criticisms derive from the inherent assumption of the development philosophy, i.e. that for greater efficiency to be achieved structural changes in the productive system will be required. The unthinking attempts in the cause of development to transfer new tech-nologies into traditional societies have often led, as we have seen, to distress and misery. The concept of change for its own sake, which is implicit in development theory, has been one of the factors bedevilling resource man-agement in the arid lands over the last century.

Changing appraisals of the Great Plains

The semi-arid Great Plains of the USA offer clear documentation of the influence of changing philosophies upon the appraisal of its resources. Examination of European and Anglo-American appraisals from the early sixteenth century onwards suggests a sequence of at least five and possibly six different regional images, most of which reflected broader philosophies of resource management.

The Great Plains presented the Europeans with a sweep of natural grass-lands almost devoid of surface water, crossed by shallow river valleys with gallery woodlands, where a fringe of indigenous agricultural villages on the more humid east merged into an area of nomadic hunters and gatherers in the drier west. The Spanish, probing north (from modern Mexico) in the early sixteenth century, found in the villages no material wealth of gold or silver, no mines of precious metals, and relatively few souls to be saved. The nomads had even less to offer potential conquerors. As a result, with Coronado's unsuccessful exploration for the fabled 'Seven Cities of Gold' (1538–42), Spanish interest evaporated. There were no resources to be exploited.

Apart from tentative fur-trapping expeditions to the northern plains in the late eighteenth century, the French interest appears to have been peripheral – since the main fur-bearing animals still lay within the wooded humid east of the continent. The 1803 Louisiana Purchase of the Great Plains by the newly independent United States from the French therefore can be explained only as part of a political strategy to acquire space – to achieve the Manifest Destiny of the new nation by occupying the land as a routeway to the markets of the 'East' (China and Japan). The purchase created a massive 'public domain' whose resources were assessed by three military expeditions: in 1803–4 Lewis and Clark via the Missouri River to the northern plains and Rocky Mountains; in 1806–7 Lieutenant Pike's group across the southern plains; and in 1819–20 Major Long's group across the central plains to present-day Colorado. While Lewis and Clark found the valleys agriculturally attractive and commented on the abundance of game, the two southern expeditions which crossed the treeless and generally waterless broad interfluves were not as impressed, despite the fact that they crossed the area in relatively moist seasons. The lack of timber for fuel and construction purposes, the lack of surface water, and the hot summers were seen as major barriers to agriculture although livestock grazing was thought possible. The area was in the words of Major Long's geographer Edwin James:

> almost wholly unfit for cultivation, and of course uninhabitable by a people depending upon agriculture for their subsistence.
>
> (Brown 1948: 372)

The Great American Desert had appeared on the map and was to dominate appraisals from the 1820s to the 1850s. Although there is dispute as to the extent and significance of this image of the plains, there is no doubt that agricultural settlement was effectively discouraged. This discouragement was possibly a political expediency, since there were fears that indiscriminate occupation of space would lead to inadequate services and the reduction of the civilizing influences such services would bring.

By the mid-nineteenth century a combination of circumstances led to a modification of the desert image. The passage of thousands of emigrants across the northern plains along the Overland Trail to Oregon from 1848 onwards, of prospective miners across the central plains *en route* to the gold-rush in California from 1849 and the gold- and silver-rushes in Colorado in the 1850s, considerably widened popular experience of the plains environment. In addition, more official expeditions such as Fremont 1842–4, Hayden 1873–4, and Powell 1867–70, added basic scientific information.

Reappraisal of the earlier desert image was general. When this was allied with political pressure to open up the land for the family farms of the 1862 Homestead Act, and with plans for transcontinental railways to allow commerce to move across the plains – railways which were to be paid for by grants of land *on* the plains – it is not surprising that the previous image was reversed and the plains came to be seen as the 'Garden' of the West (Emmons 1971). Neither settlers nor railway company directors would have appreciated free *desert* real estate. Some, particularly Powell in his 1878 *Report on the Arid Lands of the United States*, counselled caution in the use of the plains west of the 100th meridian because of their aridity, and each new drought brought bankruptcies, temporary land abandonment and rain-makers ('pluviculturalists') of all kinds – from scientists to quacks (Spence 1980). Yet the following wetter seasons brought fresh optimists and new development capital into the plains.

The Depression and Dust Bowl of the 1930s brought a new appraisal – the Great Plains were recognized to be a *problem area*, a region of environ-mental, economic, and social stress which national policies and finances had to consider (Kraenzel 1955). As noted in Chapter 12 (p. 184) and Table 16.4, since the 1930s increasing amounts of federal funds have been spent in the Great Plains, either as direct financial supports in the form of loans to farms or indirectly through soil conservation works, agricultural extension services, crop insurance indemnities, and farm price support schemes. The 1980s will see the continuation of this view with periodic droughts still bring-ing extensive economic stress. The rural population will continue to decline, although the small-town populations may benefit from both the massive coal-mining projects, developed in the northern plains in the latter half of the 1970s to offset rising oil prices and rising industrial energy demands in the eastern USA, and from new small 'footloose' industries seeking cheaper, non-union labour. For the farmer prepared to risk the vagaries of the semi-arid climate, however, the plains will continue to be attractive, not least because the terrain and soils enable continued mechanization of most agri-cultural processes and thereby favour the 'agri-business' approach to farm production.

The Idol of the Cave: the personal biases

Each of the resource managers in the arid lands has a personal view of the potentials. Such views appear explicitly in resource management and might be described as forming a matrix ranging from motives varying according to the scale, to those varying according to the type of interest of management (Table 18.2). From the individual farmer, miner, or pastoralist to the board of directors of the development company, from the local officials to the cen-tral bureaucrats, there is a wide spectrum of beliefs concerning the potentials of the arid lands, how those potentials should be realized, and who should benefit. This variety of motives and beliefs in turn not only affects human activity *in* the arid lands but also influences knowledge *about* those arid lands. Two examples, the role of the explorers and the variety of current attitudes towards desertification, illustrate the complex role of these biases.

Table 18.2 Management matrix for the arid lands

Scale of management motives	Scale of management decisions			
	Local	*Regional*	*National*	*International*
Private				
Private managers:	Farmers	⟵ Tribal groups ⟶		
individuals	Pastoralists			
groups	Miners			
companies	⟵——————— Companies ——————— (Multinationals) ⟶			
General Public:		⟵——— General public ———		⟵– – – ? – – –
unaffected				
affected				
Mixed				
Institutions:	⟵— Community Associations —⟶		National Trust	
philanthropic	⟵—— Religious groups ——⟶			⟵– – ? – –
interest groups				
Public				
Government:	Local government?	State government?	National government	UN Agencies:
executive				FAO
legislative				UNEP
administrative				etc.
judiciary				

The explorer's image

Two studies have demonstrated the role of perceived versus real knowledge in the general process of exploration and the way in which that perceived knowledge as a 'conceptual filter' influences the aims and scope of actual exploration (Allen 1972: Overton 1981). The explorer sets out with certain expectations in mind. As his journey progresses these are modified by his experiences which are interpreted personally and integrated into the explorer's report. That report is then evaluated by the sponsors of the expedition in the light of *their* expectations and prior demands.

The reports of the explorers have been evaluated also by the general public and the acceptance (and therefore the diffusion) of the explorer's personal image of the territory has varied considerably. In his study of the European exploration of the River Niger in West Africa, Lloyd (1974: 161–2) compared the popularity of the account of the British explorer, David Livingstone, with that of the German, Heinrich Barth:

> Livingstone's book sold over 50,000 copies; Barth's scarcely 2,250. Yet the two men met and exchanged signed copies in mutual admiration. Both were solitaries, travelling alone for the most part of their long sojourns in the unknown interior: but whereas Livingstone's transparent humanity and selfless zeal endeared him to the British public, Barth's strictly scientific approach to his task of delineating a new area of the continent in 3,500 pages had little appeal.

To what extent the difference reflected a popular jingoistic view of the authors or to what extent Livingstone's accounts fitted the contemporary British attitudes to Africa better than Barth's may be debated. There seems little doubt, however, that their personal views of the exploration process affected the popularity of their accounts and the diffusion of their 'humanitarian' versus 'scientific' images of Africa.

Desertification: differing perceptions of the problem

The United Nations Conference of 1977 focused attention upon the apparent deteriorating condition of the arid lands and identified the cause as desertification:

> The intensification or extension of desert conditions: it is a process leading to reduced biological productivity, with consequent reduction in plant biomass, in the land's carrying capacity for livestock, in crop yields and human wellbeing. (UN 1977a: 3)

The areas affected were mapped and recommendations for amelioration of conditions proposed (Fig. 18.2 and Table 18.3). Five years later the process continued relatively unabated and part of the lack of success in controlling it seemed to lie in the differing perceptions of desertification and the threat it poses (Glantz 1980; Walls 1980).

Fig. 18.2 Desertification in the arid lands. *Source*: UN 1977a

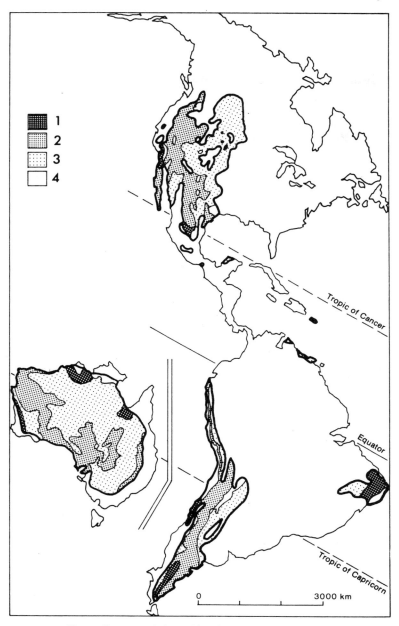

Key: Degree of desertification hazard:
1 – Very high; 2 – High; 3 – Moderate; 4 – None.
Note that the desertification hazard is only mapped within the arid lands.

Table 18.3 Estimates of global desertification

A. Land affected, cost of and net gains from rehabilitation

	Total area affected (million ha)	Annual rate of land degradation (000 ha)	Cost of salvage $million	Net gain from salvage $million
Irrigated Waterlogged	25	125	106	112.5
Salinity problems	20			
Rangeland	3600	3600	180	180.0
Rain-fed cropland	170	1700	170	340.0
Total	3815	5425	456	632.5

B. Population living in areas recently undergoing severe desertification

	Area affected (million km²)	Percentage of population by livelihood type			Total population (million)
		Urban based	Cropping based	Animal based	
Mediterranean	1.3	31	60	9	9.8
Sub-Saharan Africa	6.9	19	37	44	16.2
Asia and the Pacific	4.4	27	54	19	28.5
Americas	17.6	32	56	12	24.1
Total	30.1	27	51	22	78.6

Source: UN Secretariat 1977.

Studies have identified a variety of knowledge of and attitudes to desertification (Hare 1980; Heathcote 1980). From the scientists to the politicians to the farmers there were different images. While the scientists debated its character and extent they generally agreed that it was a serious problem for mankind, especially in the areas affected. The politicians saw it as a convenient stick to beat the government – if in opposition, or to raise foreign aid – if in government. The majority of farmers saw the problem as either tolerable or one over which they had no control – a kind of occupational hazard about which they could do nothing, and a problem overshadowed by more pressing stresses from rising costs and falling incomes. For a minority, indeed, the hazard was not recognized at all, either because of ignorance, an unwillingness to accept faults in their own land, or the belief that it was a natural process in which human activity had played at most only a minor role.

Any attempts to combat desertification will need to be aware of this variety of images. Scientists need to improve their understanding of the process as the interplay between environmental characteristics and human activities, and to recognize that their 'objective' view is based upon a relatively secure job and income and is an *outsider's* view. Governments will need both to convince the 'front-line troops' – the farmers – that desertification *is* a problem, as well as provide them with the means and the incentives to fight it. At the same time governments will need to recognize the many *other* pressing problems facing the farmers and provide not only efficient but sympathetic advisory services.

The Idol of the Tribe: traditional systems

Traditional systems of resource management stem from both traditional beliefs and value-systems expressed as a society's culture and from the traditional institutions set up to enforce those beliefs. Two examples from the arid lands illustrate the role of both influences on resource management.

Culture clash: conservation – no good for Navajos?

Fonaroff (1963) documented a classic culture clash on the arid rangelands of the Navajo Reservation in the southwest USA. The clash was between two different perceptions of the nature and significance of soil erosion on the range. On the one hand was the Anglo-American view – expressed through the Bureau of Indian Affairs responsible for the administration of the reserve – that the excessive soil erosion, evident by the 1930s, could be stopped by reduction of livestock numbers to ease grazing pressure on the range, and by engineering works on the catchments and drainage lines. On the other hand, the Navajo saw soil erosion as a natural phenomenon unaffected by grazing pressures. The gullies had always been there and people were powerless to influence them. Furthermore, livestock were important to the Navajo as a source of wealth, social status, and security. Any attempts to reduce their flocks, especially by the apparently senseless slaughter of excessive numbers, were vigorously resisted. At the time Fonaroff wrote, 30 years after the preparation of an official soil conservation plan for the reserve, sheep numbers were still too large, there was still resistance to any rotational use of the range and soil erosion was still extensive. In 1978 field observation confirmed that soil erosion was still evident. Conservation, according to their own view, was no good for Navajos.

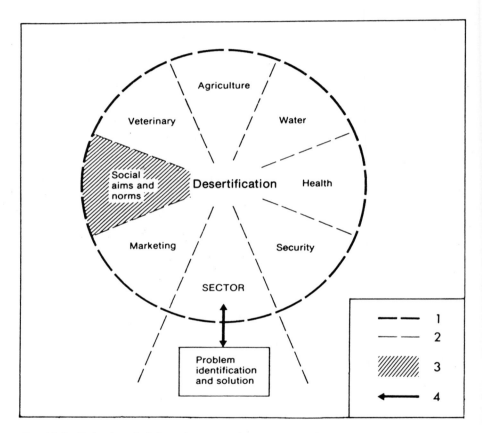

Fig. 18.3 Baker's 'administrative trap'. *Source*: Baker 1976b

    Key: 1 – Problem boundary (ecologically determined)

           2 – Symptom boundary (sectoral division of responsibility in government, departmental boundary). Lack of communication across these boundaries.

           3 – Sector 'largely neglected'

           4 – Axis of problem 'solution' by sectors

Bureaucracy: the 'administrative trap'

In an analysis of apparently inadequate official responses to the famines in the Sahel between 1968 and 1974 Baker (1976a, 1976b) suggested that part of the problem was the method by which the problem was tackled by the various national and international bureaucracies. Despite the fact that the ramifications of the famine affected the whole of the various national environments and economies, the government departments tended to tackle only that portion of the problem which fell within their jurisdiction. As a result the overall problem was tackled only by sectors which were never co-ordinated at any scale (Fig. 18.3). As a result of this lack of co-ordination, an 'administrative trap' was created, which prevented effective mitigation policies being implemented even if they had been available.

While there is debate about the origins of official bureaucracies, there is no doubt that the scope of their various jurisdictions has become wider and more complex since the nineteenth century. This has been the result of technological innovations which have created complex systems of communication, trade, and production which are thought to need national supervision because of their national relevance. In addition, political ideologies such as socialism and communism have emphasized the need for government management of national resources for the benefit of the general public – management in effect by the bureaucratic civil servants. As a result, separate government departments for agriculture, mines, water supplies, transport, and public health, have become standard sectors of official concern. The existence of regional co-ordinating systems of government – suggested by Baker as a way out of the administrative trap – is rare and usually only exists in an advisory capacity. Here, it is the traditional bureaucratization of government which blinkers both the perception of and reactions to the problems of resource management. This sectoral view ignores the regional distinctiveness of many resource-management problems and favours isolated uncoordinated projects, not regional and comprehensive planning.

CHAPTER 19

Conclusions

The global arid lands constitute one of the world's major ecosystems and provide a significant contribution to the global economy. Covering one-third of the world's land area, they provide at least a fifth of the world's food supplies, over half the world's production of precious and semi-precious minerals, and the bulk of the oil and natural gas production and reserves. Yet only approximately 14–15 per cent of the world's population lives in the arid lands, and those mainly on the semi-arid fringes, while their condition varies from high levels of material affluence to propertyless poverty.

A highly variable climate has had marked impacts upon human activities. Yet those activities themselves have imposed demands upon the arid eco-system, demands which in many cases have been beyond the capacity of the ecosystem to supply. As a result, the history of human activity in the arid lands is marked by both evidence of magnificent achievements and catas-trophic failures. There is the long history of the irrigated lands where the 'contriving brain and skilful hand' of man transformed disease-ridden swamps or 'barren sands' into resources to support some of the highest popu-lation densities ever experienced in the world. Associated with those con-centrations of population has been the creation of surpluses to support creative thought, philosophies, cultures, and artefacts, the latter still includ-ing some of the wonders of the world.

But it is perhaps in the nature of humanity that against that history of creative effort and skilful manipulation of resources to human advantage we must set the equally long history of human conflict in the arid lands; conflict which led to the wilful destruction of those very artefacts so wondrously produced. In the arid lands of the 1980s not only are there massive contrasts in standards of living among the peoples, but also massive differences in their use of the land – from the subsistence hunter–gatherers to the agri-business wheat firms, from terrains whose only intrusion is the click of tourist cameras to terrains eroding from over-use and those still radioactive from past and continuing nuclear weapon tests.

While it is as easy to condemn the massive investments in armaments as to condemn the misuse of arid lands resources, it is just as difficult to foster optimal resource use as it is to beat swords into ploughshares. The years since the end of the Second World War have seen many attempts to improve the productivity of the arid lands. The dearly bought conclusion, recognized by the late 1970s, was that intensification of production would be limited in space and not necessarily continuous through time. While there was no doubt that significant successes had been achieved, it was equally clear that much money and effort had been wasted, that many of the 'development projects' had led either to a deterioration of the condition of both the land and its inhabitants or merely accelerated a rural depopulation that seemed inevitable. Further, the effect of development has been to concentrate human activities upon smaller areas of the arid lands, leaving once produc-tive areas unused – not necessarily because their productivity had decreased, although desertification was widespread, but because they no longer offered a life style sufficiently attractive to retain the population.

Faced by this evidence of apparent failure to plan for improved resource management, faced by the increasing human population in the arid lands, and increasing evidence of the degradation of the pastoral and agricultural resources, what guidance can be offered to the future managers of the arid lands? Five points are suggested.

First, the managers will need to recognize that their number will be large and that each may have equally important, useful, rational, but probably *different* strategies for resource management. The weighting of these alterna-tives, the necessary consultations and compromises, and the sensitive under-standing of contrasting proposals is the basic task of the future managers.

Second, they should be aware of the historical geography of human activ-ity in the arid lands. There is much to be discovered about and learned from the sequence and changing patterns of human settlements in the arid lands. While George Bernard Shaw's conclusion may have been true – that 'we learn from experience that we do not learn from experience' – this can be no excuse for the future managers who already have in the literature (sci-entific, official, and popular) and oral history a wealth of human experience to draw upon.

Third, the bulk of the historical evidence suggest that in the long term – more than a single generation – the resource-use systems which have endured have been those where management has been sufficiently flexible to accommodate the fluctuations in resource availability. Surpluses from the good years have been retained to carry the society over the bad years, or the scale of management has been broad enough to allow regional deficits in one place to be compensated by regional surpluses in another. Entire reliance upon only one resource strategy usually has proved disastrous – whether for ecological or economic reasons. In such cases short-term local benefits have usually been more than outweighed by long-term costs.

Fourth, depending upon their political ideology, the resource managers

297

may find the concept of planning an essential component of management or an anathema. The existence of elaborate national planning structures is no guarantee of good planning, even if planning is defined as resource management for the maximum benefit of the greatest number. What may be necessary is a balance between traditional and innovative management strategies and individual initiative and societal controls. A rejection of any innovations will expose that arid lands society to severe social pressures which might eventually destroy that society from within: unthinking acceptance of innovations just because they are innovations will destroy the society in any case. The contemporary revival of fundamentalist Muslim ideologies in Iran would be a possible example of the first situation, while the stresses among the Sahel nations seem to illustrate the second. In contrast, unlimited individual resource exploitation may produce private gain at considerable public expense: totalitarian state planning will stultify individual initiative and innovation and produce development schemes which may be grandiose failures. Settlement of the arid lands occupied by Western Europeans since the mid-nineteenth century offers abundant examples of the first situation, while the Russian Virgin Lands scheme disaster of the 1950s provides an example of the second. In the experience of the arid lands no political ideology has yet provided the optimal management strategy.

Finally, in the future it may be that each of the conflicting ideologies and associated management systems may have a role to play. The traditional resource-use systems offer invaluable experience of management with limited resources but tangible successes. From individual initiatives will come tested innovations to apply capital, skills, and labour more efficiently; while centralized planning agencies will allocate scarce resources between competing potential users. The role of the managers will be to concoct the best mix of these factors and apply them to the arid lands.

There are, indeed, some indications that future management of the arid lands will be more sensitive to the variety of the possible management strategies. In part this will result from the recognition by the external development agencies of the need to accommodate local aspirations and preferred developments; to concentrate upon smaller rather than larger projects; to encourage self-help rather than projects entirely dependent upon foreign expertise dispensed by temporarily seconded 'experts'; and to favour developments using minimal fossil fuel energy. Such trends will be favoured by the increasing wealth and influence of some of the core and predominantly arid nations – wealth based, ironically, upon oil revenues and influence based upon the revival of the Muslim faith and its associated cultural values. Increasingly, such nations may be able to instigate and control their own resource management rather than merely responding to the proposals from outsiders.

Currently, however, the arid lands contain some of the poorest nations in the world and large areas whose productivity is rapidly declining through the combined pressures of rapidly increasing human and animal populations.

They also contain some of the largest areas of solar energy potential, and of spacious and relatively undisturbed natural environments as yet unpolluted by industrial wastes. Over some of the oldest battlefields of the world manoeuvre some of the most sophisticated armaments of the modern world: close by, on some of the oldest farmlands in the world are some of the most sophisticated irrigation aids to farming yet produced by man. The irony and potential menace of such a situation is the fundamental challenge to future resource management in the arid lands.

UNESCO's Arid Zone Research Series

    I. *Reviews of Research on Arid Zone Hydrology*, 1953, 212 pp.

   II. *Proceedings of the Ankara Symposium on Arid Zone Hydrology*, 1953, 268 pp.

  III. *Directory of Institutions Engaged in Arid Zone Research*, 1953, 110 pp.

  IV. *Utilization of Saline Water. Reviews of Research*, 1954, 96 pp.

   V. *Plant Ecology*. Proceedings of the Montpellier Symposium, 1955, 124 pp.

  VI. *Plant Ecology. Reviews of Research*. 1955, 377 pp.

 VII. *Wind and Solar Energy*. Proceedings of the New Delhi Symposium, 1956, 238 pp.

VIII. *Human and Animal Ecology. Reviews of Research*, 1957, 244 pp.

  IX. *Guidebook to Research Data on Arid Zone Development*, 1957, 191 pp.

   X. *Climatology. Reviews of Research*, 1958, 190 pp.

  XI. *Climatology and Microclimatology*. Proceedings of the Canberra Symposium, 1958, 355 pp.

 XII. *Arid Zone Hydrology. Recent Developments*, 1959, 125 pp.

XIII. *Medicinal Plants of the Arid Zones*, 1960, 96 pp.

XIV. *Salinity Problems in the Arid Zones*. Proceedings of the Teheran Symposium, 1961, 395 pp.

 XV. *Plant – Water Relationships in Arid and Semiarid Conditions. Reviews of Research*, 1960, 225 pp.

XVI. *Plant – Water Relationships in Arid and Semiarid Conditions*. Proceedings of the Madrid Symposium, 1962, 352 pp.

XVII. *A History of Land Use in Arid Regions*, 1961, 388 pp.

XVIII. *The Problems of the Arid Zone*. Proceedings of the Paris Symposium, 1962, 481 pp.

XIX. *Nomades et Nomadisme au Sahara*, 1963, 195 pp.

 XX. *Changes of Climate*. Proceedings of the Rome Symposium organized by UNESCO and WMO, 1963, 488 pp.

XXI. *Bioclimatic Map of the Mediterranean Zone. Explanatory Notes*, 1963, 58 pp.

XXII. *Environmental Physiology and Psychology in Arid Conditions. Reviews of Research*, 1963, 345 pp.

XXIII.   *Agricultural Planning and Village Community in Israel*, 1964, 159 pp.
XXIV.   *Environmental Physiology and Psychology in Arid Conditions*. Proceedings of the Lucknow Symposium, 1964, 400 pp.
XXV.   *Methodology of Plant Ecophysiology*. Proceedings of the Montpellier Symposium. 1965, 531 pp.
XXVI.   *Land Use in Semiarid Mediterranean Climates*, 1964, 170 pp.
XXVII.   *Evaporation Reduction, Physical and Chemical Principles and Review of Experiments*, 1965, 79 pp.
XXVIII.   *Geography of Coastal Deserts*, 1966, 140 pp.
XXIX.   *Physical Principles of Water Percolation and Seepage*, 1968, 465 pp.

References

AAG (1980) Assoc. Amer. Geogs., Washington, *Newsletter*, **15** (7).

Adams, R. M. (1965) *Land behind Baghdad: a history of settlement on the Diyala Plains*, University of Chicago, Chicago.

Adams, W. P. and Helleiner F. M. (eds) (1972) *International Geography 1972*, Vol. 2, University of Toronto Press, Toronto.

Adams, R. and M., and Willens A. and A. (1978) *Dry Lands: man and plants*, Architectural Press, London.

Aizenshtat, B. A. (1966) 'Investigations of the heat balance of Central Asia' in M. I. Budyko (ed.) *Contemporary Problems of Climatology*, Gidrometeoizdat, Leningrad.

Akramov, Z. M. (1976) 'Interbranch and territorial production complexes in desert regions (with reference to Central Asia)', *Int. Geog. Congress Moscow*, **6**, 76–80.

Allan, J. A. (1976) 'The Kufrah agricultural schemes', *Geog. J.*, **142**, 50–6.

Allan, W. (1965) *The African Husbandman*, Oliver and Boyd, Edinburgh.

Allen, J. A. (1972) 'An analysis of the exploratory process: the Lewis and Clark Expedition of 1804–1806', *Geog. Rev.*, **62**, 13–39.

Amiran, D. H. K. (1978) 'Geographical aspects of national planning in Israel: the management of limited resources', *Trans. Inst. Brit. Geogs.*, New Series, **3**, 115–28.

Amiran, D. H. K. and Ben-Arieh, Y. (1963) 'Sedentarization of Bedouin in Israel', *Israel Exploration J.*, **13**, 161–81.

Amiran, D. H. K. and Ben-Arieh, Y. (eds) (1976) *Geography in Israel*, Israel Nat. Comm. Geog., Jerusalem.

Amiran, D. H. K. and Wilson, A. W. (eds.) (1973) *Coastal Deserts: their natural and human environments*, University of Arizona, Tucson.

Andreae, B. (1966) 'Weidewirtschaft im südlichen Afrika', *Geog. Zeitschift Beihefte Erdkundliches Wissen-Heft*, **15**, Wiesbaden.

Arifi, S. A. El- (1979) 'Some aspects of local government and environmental management in the Sudan' in Mabbutt (ed.) 1979: 36–9.

Armillas, P. (1961) 'Land use in pre-Columbian America' in Stamp (ed) 1961: 255–76.

Atherton, L. (1961) *The Cattle Kings*, Indiana University, Bloomington.

Australia (1978) *A Basis for Soil Conservation Policy in Australia*, Dept. Environment, Housing and Community Devt., Res. Directorate Report No. 1, Canberra.

Awad, H. (1963) 'Some aspects of the geomorphology of Morocco related to the Quaternary climate', *Geog. J.*, **129**, 129–39.

Awad, H. (1971) 'The oasis towns of the Sahara', *Selected Papers of 21st Int. Geog. Congress*, Calcutta, **3**, 12–21.

Ayensu, E. S. (1979) 'Plants for medicinal uses with special reference to arid zones' in Goodin and Northington (eds), 1979: 117–78.

Bagnold, R. A. (1941) *The Physics of Blown Sand and Desert Dunes*, Methuen, London.

Baker, R. (1976a) 'Innovation technology transfer and nomadic pastoral societies' in Glantz (ed.) 1976b: 176–85.

Baker, R. (1976b) 'The administrative trap', *The Ecologist*, **6**, 247–51.

Baker, R. St Barbe (1966) *Sahara Conquest*, Lutterworth, London.

Baumhoff, M. A. (1978) 'Review of Cohen 1977' in *Agric. Hist.* **5**, 770–2.

Beaumont, P. (1974) 'Water resource development in Iran', *Geog. J.*, **140**, 418–31.

Bedoian, W. H. (1978) 'Human use of the pre-Saharan ecosystem and its impact on desertization' in Gonzalez (ed.) 1978: 61–109.

Belgrave, J. H. D. (1975) *Welcome to Bahrain*, Augustan Press, Manama.

Bennett, H. H. (1947) *Elements of Soil Conservation*, McGraw-Hill, New York.

Bennett, R. J. and Chorley R. J. (1978) *Environmental Systems: philosophy analysis and control*, Methuen, London.

Berque, J. (1959) 'Nomads and nomadism in the Arid Zone', *Int. Soc. Sci. J.*, **11**, 481–98.

Bhimaya, M. (1960) 'Comment' in *Arid Zone Research Series No. 18*, UNESCO, Paris, 1960: 365.

Black, J. N. (1956) 'The distribution of solar radiation over the earth's surface' in *Wind and Solar Energy, Arid Zone Research Series No. 7*, UNESCO, Paris.

Blainey, G. (1969) *The Rush That Never Ended*, Melbourne University Press, Carlton (1st edn, 1963).

Bobek, H. (1962) 'The main stages in socio-economic evolution from a geographical point of view' in Wagner and Mikesell 1962: 218–47.

Bonine, M. E. (1979) 'The morphogenesis of Iranian cities', *Annals Assoc. Amer. Geogs.*, **69**, 208–24.

Borchert, J. R. (1971) 'The Dust Bowl in the 1970s', *Annals Assoc. Amer. Geogs.*, **61**, 1–22.

Bourgeois, J. L. (1980) 'Welcoming the wind', *Natural History*, **11**, 70–5.

Bowman, I. (1924) *Desert Trails of Atacama*, Amer. Geog. Soc., New York.

Box, T. W. (1973) 'The future use of Australia's arid lands' in University of Adelaide 1973: 17.1–17.10.

Boyko, H. (ed.) (1966) *Salinity and Aridity: new approaches to old problems*, Junk, Hague.

Braidwood, R. J. (1971) 'The agricultural revolution' in Ehrlich (ed.) 1971: 17–25.

Brazel, A. J. and Idso, S. B. (1979) 'Thermal effects of dust on climate', *Annals Assoc. Amer. Geogs.*, **69**, 432–37.

Briggs, J. A. (1978) 'Farmers' responses to planned agricultural development in the Sudan', *Trans. Inst. Brit. Geogs.*, New Series, **3**, 464–75.

Brittan, M. (1979) *Discover Namibia*, Struik, Cape Town.

Bromehead, C. N. (1956) 'Mining and quarrying' in Singer *et al.* (eds) 1956: Vol. I, 558–71.

References

Brown, R. H. (1948) *Historical Geography of the United States*, Harcourt Brace, New York.

Bryson, R. A. and Baerries, D. A. (1967) 'Possibilities of major climatic modification and their implications: northwest India, a case for study', *Bull. Amer. Meteor. Soc.*, **48**, 136–42.

Butzer, K. (1980) 'Civilizations: organisms or systems?', *Amer. Scientist*, **68**, 517–23.

Cain, A. *et al.* (1976) 'Traditional cooling systems in the Third World', *The Ecologist*, **6**, 60–4.

Calder, R. (1951) *Men Against the Desert*, Allen and Unwin, London.

Cameron, R. J. (comp.) (1979) *Year Book Australia No. 63, 1979*, Australian Bureau of Statistics, Canberra.

Caponera, D. A. (1973) *Water Laws in Moslem Countries*, Irrigation and Drainage Paper No. 20/1, FAO, Rome.

Carr, M. H. (1980) 'The geology of Mars', *Amer. Scientist*, **68**, 626–35.

Carrier, E. H. (1932) *Water and Grass: a study in the pastoral economy of southern Europe*, Christophers, London.

Carvalho, E. C. de (1974) ' "Traditional" and "Modern" patterns of cattle raising in southwestern Angola: a critical evaluation of change from pastoralism to ranching', *J. Developing Areas*, **8**, 199–226.

Childe, G. (1952) *What Happened in History*, Penguin, Harmondsworth. (1st edn, 1942).

Childe, V. G. (1951) *Man Makes Himself*, Mentor, New York.

Chippendale, G. M. (1968) 'A study of the diet of cattle in central Australia as determined by rumen samples', *Northern Territory Primary Industries Branch Techn. Bull.* **1**, Darwin.

Chorley, R. J. *et al.* (eds) (1969) *Water, Earth and Man*, Methuen, London.

Christaller, W. (1966) *Central Places in Southern Germany*, (trans. C. W. Baskin), Prentice-Hall, Eaglewood Cliffs.

Christopher, A. J. (1976) *Southern Africa*, Dawson, Folkestone.

Clawson, M. (1963) 'Recreation as a competitive segment of multiple use' in Thorne (ed.) 1963: 169–83.

Clawson, M., Held, R. B. and Stoddard, C. H. (1960) *Land for the Future*, Johns Hopkins, Baltimore.

Cloudsley-Thompson, J. L. (1965) *Desert Life*, Pergamon, Oxford.

Cloudsley-Thompson, J. L. (1969) 'Desert' in *The Encyclopaedia Americana*, Vol. 9, 1–10.

Cloudsley-Thompson, J. L. and Chadwick, M. J. (1964) *Life in Deserts*, Foulis, London.

Cohen, M. N. (1977) *The Food Crisis in Prehistory: overpopulation and the origins of agriculture*, Yale University, New Haven.

Coombs, H. C. (1979) 'Science and technology for what – or for whom?', *Current Affairs Bull.* (Sydney), **56**, 4–15.

Coon, C. S. (1971) *The Hunting Peoples*, Little, Brown, Boston.

Cordes, R. and Scholz, F. (1980) *Bedouins, Wealth and Change*, United Nations University, Tokyo.

Crossley, J. C. (1976) 'The location of beef processing', *Annals Assoc. Amer. Geogs.*, **66**, 60–75.

CSIRO (1979) 'Growing fuel on the farm', *Rural Research*, **103**, 4–11.

Cunha, E. da (1957) *Rebellion in the Backlands*, (trans. S. Putnam), University of Chicago, Chicago (1st edn, 1902).

Daniels, F. (1967) 'Direct use of the sun's energy', *Amer. Scientist*, **55**, 15–47.

Daryll Forde, C. (1950) *Habitat, Economy and Society*, Methuen, London, (1st edn., 1934).

Das Gupta, S. P. (ed.) (1971) *Proceedings of Symposium on Arid Zone*, Indian National Committee for Geography, Calcutta.

Davies, H. R. J. (1966) 'Nomadism in the Sudan: aspects of the problem and suggested lines for its solution', *Tijd. Econ. Soc. Geog.*, **57**, 193–202.

Davis, D. H. S. (ed.) (1964) *Ecological Studies in Southern Africa*, Mon. Biol., **XIV**, Jung, Hague.

Davis, W. M. (1905) 'The Geographical Cycle in an arid climate', *J. Geology*, **13**, 381–407.

Davitaya, F. F. (1969) 'Atmospheric dust content as a factor affecting glaciation and climatic change', *Annals Assoc. Amer. Geogs.*, **59**, 552–60.

Dawood, N. J. (trans.) (1956) *The Koran*, Penguin, Harmondsworth.

Deevey, E. S. (1971) 'The human population' in Ehrlich (ed.), 1971: 48–55.

Denevan, W. M. (1967) 'Livestock numbers in 19th Century New Mexico and the problems of gullying', *Annals Assoc. Amer. Geogs.*, **57**, 691–703.

Dick, J. and Marden, C. J. (1979) 'What is the basis of current energy forecasts?' in Diesendorf (ed.), 1979: 7–19.

Diesendorf, M. (ed.) (1979) *Energy and People*, Soc. Social Responsibility in Science, Canberra.

Dolan, R., Howard, A. and Gallenson, A. (1974) 'Man's impact on the Colorado River in the Grand Canyon', *Amer. Scientist*, **62**, 392–401.

Dregne H. E. (ed.) (1970) *Arid Lands in Transition*, Amer. Assoc. Adv. Sci., Pub. No. 90, Washington.

Dregne, H. E. (1976) *Soils of the Arid Regions*, Elsevier, Amsterdam.

Dresch, J. (1966) 'Utilization and human geography of the deserts', *Trans. Inst. Brit. Geogs.*, **40**, 1–10.

Drower, M. S. (1956) 'Water-supply, irrigation and agriculture' in Singer *et al.* (eds) (1956): Vol. I, 520–57.

Duly, L. C. (1968) *British Land Policy at the Cape, 1795–1844*, Duke University, Durham.

Dunbier, R. (1968) *The Sonoran Desert*, University of Arizona, Tucson.

Duwaji G. (1968) 'Land ownership in Tunisia: an obstacle to agricultural development', *Land Economics*, **44**, 129–32.

Ehrlich, P. R. (ed.) (1971) *Man and Ecosphere* (Readings from *Scientific American*), Freeman, San Francisco.

Eldridge, F. E. (1976) *Wind Machines*, Nat. Sci. Fdn., Washington.

Elgabaly, M. M. (1977) 'Salinity and waterlogging in the Near-East Region', *Ambio*, **6**, 36–39.

Emmons, D. M. (1971) *Garden in the Grasslands*, University of Nebraska, Lincoln.

Erickson, E. W. and Waverman, L. (1974) *The Energy Question: an international failure of policy*, Vol. 1. *The World*, University of Toronto, Toronto.

Evans, J. V. (1979) 'The potential for desalinating sea water' in Hallsworth, E. G. and Woodcock, J. T. (eds) *Land and Water Resources in Australia*, Aust. Acad. Tech. Sciences, Parkville, 133–58.

References

Evenari, M., Shanan, L. and Tadmor, N. (1971) *The Negev: the challenge of a desert*, Harvard University, Cambridge.

Fawcett, R. G. (1978) *Research Needs in the Murray Mallee*, South Australian Dept. Agric: and Fisheries, Adelaide.

Field, J. (ed.) (1964) *Handbook of Physiology. Section 4: Adaptations to the Environment*, Amer. Physiological Soc., Washington.

Fink, D. H. and Ehrler, W. L. (1979) 'Runoff farming for jojoba' in Goodin and Northington 1979: 212–24.

Flannery, K. V. (1971) 'Origins and ecological effects of early domestication in Iran and the Near East' in Ucko and Dimbleby (eds) 1971: 73–100.

Fonaroff, L. S. (1963) 'Conservation and stock reduction on the Navajo tribal reserve', *Geog. Review*, **53**, 200–23.

Fowler, G. L. (1972) 'Italian colonization of Tripolitania', *Annals Assoc. Amer. Geogs.*, **62**, 627–40.

Fowler, G. L. (1973) 'Decolonization of rural Libya', *Annal Assoc. Amer. Geogs.*, **63**, 490–506.

Freeberne M. (1966) 'Demographic and economic changes in Sinkiang vighur autonomous region', *Population Studies*, **20**, 103–24.

Frobel, F., Hendricks J. and Kruge, D. (1979) *The New International Division of Labour*, Cambridge University Press, Cambridge.

Gartside, G. (1977) 'The energy cost of prospective fuels', *Search*, **8**, 105–10.

Gates, P. W. (1968) *History of Public Land Law Development*, US Govt. Print. Office, Washington.

Gautier, E. F. (1970) *Sahara: the great desert*, (trans. D. F. Mayhew), Octagon Books, New York.

Geiger, R. (1965) *The Climate Near the Ground*, (trans. Scripta Technicalno), Harvard University, Cambridge, Mass.

Gentilli, J. (1971) *Climates of Australia*, reprint of Chapters 4–7, *World Survey of Climatology*, Vol. 13, Elsevier, Amsterdam.

Gersmehl, P. J. (1976) 'An alternative biogeography', *Annals Assoc. Amer. Geogs.*, **66**, 223–41.

Gideon, S. (1948) *Mechanization Takes Command*, Oxford University, New York.

Gilbraith, K. (1973) *Red Capitalism: an analysis of the Navajo economy*, University of Oklahoma, Norman.

Glantz, M. H. (1976a) 'Nine fallacies of natural disaster' in Glantz (ed.) 1976b: 3–24.

Glantz, M. H. (ed.) (1976b) *The Politics of Natural Disaster: the case of the Sahel Drought*, Praeger, New York.

Glantz, M. H. (ed.) (1977) *Desertification: environmental degradation in and around arid lands*, Westview, Boulder.

Glantz, M. H. (1980) 'Man, state and the environment: an inquiry into whether solutions to desertification in the West African Sahel are known but not applied', *Can. J. Devt. Studies*, **1**, 75–97.

Glueck, N. (1959) *Rivers in the Desert*, Weidenfeld and Nicolson, London.

Golanyi, G. (ed.) (1979) *Arid Zone Settlement Planning: the Israeli experience*, Pergamon, New York.

Golding, E. W. (1953) 'The utilization of wind power in desert areas' in *Desert Research*, Res. Council of Israel Spec. Pub. No. 2, Jerusalem, 1953: 592–604.

Gonzalez, N. L. (ed.) (1978) *Social and Technological Management in Dry Lands: past and present, indigenous and imposed*, Westview, Boulder.

Goodall, D. W. and Perry, R. A. (1979) *Arid Land Ecosystems*, Vol. 1, Cambridge University, London.

Goodin, J. R. and Northington, D. K. (eds) (1979) *Arid Land Plant Resources*, Int. Centre Arid Land Studies, Lubbock, Texas.

Griffin, G. F. and Lendon, C. (1979) *A Report on Three Visits Through Three Aboriginal Homelands in Central Australia*, CSIRO Div. Land Resources Management, Perth.

Grove, A. T. (1977) 'The geography of semi-arid lands', *Phil. Trans. R. Soc. London*, B, **278**, 457–75.

Grove, A. T. (1978) 'Late Quaternary climatic change and the conditions for current erosion in Africa', *Geo-Eco-Trop.*, **2**, 291–300.

Grunebaum, E. von (1955) *Islam: essays on the nature and growth of a cultural tradition*, Barnes and Noble, New York.

Guelke, L. (1976) 'Frontier settlement in early Dutch South Africa', *Annals Assoc. Amer. Geogs.*, **66**, 25–42.

Gulhati, N. D. (1955) *Irrigation in the World: a global view*, Int. Comm. Irrigation and Drainage, New Delhi.

Hadley, R. F. (1977) 'Evaluation of land-use and land-treatment practices in semi-arid western United States', *Phil. Trans. R. Soc. London*, B, **278**, 543–54.

Hall, E. A. A., Specht, R. L. and Eardley, C. M. (1964) 'Regeneration of the vegetation on the Koonamore Vegetation Reserve, 1926–1962', *Aust. J. Bot.*, **7**, 205–64.

Hamdan, G. (1961) 'Evolution of irrigated agriculture in Egypt' in Stamp (ed.) 1961: 119–42.

Hamming, E. (1958) 'Water legislation', *Economic Geog.*, **34**, 42–6.

Hanson, G. P. *et al.* (1979) 'Guayule – a potential rubber crop for semi-arid lands' in Goodin and Northington (eds) 1979: 195–211.

Hare, F. K. (1980) 'The planetary environment: fragile or sturdy?', *Geog. J.*, **146**, 379–95.

Hargreaves, M. W. (1957) *Dry Farming on the Northern Great Plains*, Harvard University, Cambridge, Mass.

Harris, D. R. (1966) 'Recent plant invasions in the arid and semi-arid southwest of the United States', *Annals Assoc. Amer. Geogs.*, **56**, 408–22.

Hastings, J. R. and Turner, R. M. (1965) *The Changing Mile*, University of Arizona, Tucson.

Heady, H. F. (1975) *Rangeland Management*, McGraw-Hill, New York.

Heathcote, R. L. (1965) *Back of Bourke*, Melbourne University Press, Carlton.

Heathcote, R. L. (1967) 'The effects of past droughts on the national economy' in *Report of the ANZAAS Symposium on Drought*, Commonwealth Bur. Met. (Australia), Melbourne, 1967, 27–45.

Heathcote, R. L. (1969) 'Land tenure systems: past and present' in Slatyer and Perry (eds) 1969: 185–208.

Heathcote, R. L. (1975) *Australia*, Longman, London.

Heathcote, R. L. (1976) 'Early European perceptions of the Australian Landscape: the first hundred years' in G. Seddon and M. Davis (eds.) *Man and Landscape in Australia*, Australian Govt. Pub. Service, Canberra.

References

Heathcote, R. L. (1977) 'Pastoral Australia' in Jeans (ed.) 1977: 252–88.

Heathcote, R. L. (ed.) (1980) *Perception of Desertification*, United Nations University, Tokyo.

Heathcote,· R. L. and Twidale, C. R. (1969) *An Introduction to the Arid Lands*, Longman, Australia, Croydon.

Herbert, F. (1965) *Dune*, New English Library, London.

Hewes, L. (1965) 'Causes of wheat failure in the dry farming region, central Great Plains, 1939–1957', *Econ. Geog.*, **41**, 313–50.

Hewes, L. (1973) *The Suitcase Farming Frontier*, University of Nebraska, Lincoln.

Hewes, L. and Schmieding, A. C. (1956) 'Risk on the central Great Plains: geographical patterns of wheat failure in Nebraska, 1931–52', *Geog. Rev.*, **42**, 375–87.

Highsmith, R. M. (1965) 'Irrigated lands of the world', *Geog. Rev.*, **55**, 382–89.

Hilgard, E. W. (1906) *Soils, their Function, Properties, Composition and Relations to Climate and Plant Growth in the Humid and Arid Regions*, Macmillan, New York.

Hills, E. S. (ed.) (1966) *Arid Lands: a geographical appraisal*, Methuen, London.

Hodge, C. (ed.) (1963) *Aridity and Man: the challenge of the arid lands in the United States*, Amer. Assoc. Adv. Sci. Pub. No. 74, Washington.

Hodgson, R. E. (1963) 'Livestock interests in public and private lands of the west' in Thorne (ed.) 1963: 73–81.

Hollis, G. E. (1978) 'The falling levels of the Caspian and Aral seas', *Geog. J.*, **144**, 62–80.

Holz, R. K. (1968) 'The Aswan High Dam', *Prof. Geog.*, **20**, 230–237.

Houerou, H. N. Le (1970) 'North Africa: past, present, future' in Dregne (ed.) 1970: 227–78.

ICASALS (1968) International Center for Arid and Semi-Arid Land Studies, Lubbock, Texas, *Newsletter*, **2**(1), 6.

ICASALS (1978) International Center for Arid and Semi-Arid Land Studies, Lubbock, Texas, *Newsletter*, **11**(4), 5.

Issawi, C. (1969) 'Economic change and urbanization in the Middle East', in Lapidus (ed.) 1969: 102–21.

Jacks, G. V. and Whyte, R. O. (1938) *The Rape of the Earth: a world survey of soil erosion*, Faber and Faber, London.

Jackson, W. A. D. (1956) 'The Virgin and Idle Lands of western Siberia and northern Kazakhstan: a geographical appraisal', *Geog. Rev.*, **46**, 1–19.

Jacobsen, T. and Adams, R. M. (1958) 'Salt and silt in ancient Mesopotamian agriculture', *Science*, **128**, 1251–58.

Jeans, D. N. (ed.) (1977) *Australia: a geography*, Sydney University, Sydney.

Johnson, D. L. (1969) *The Nature of Nomadism*, University of Chicago, Dept. Geog. Res. Paper 118, Chicago.

Johnson, D. L. (1979) 'Management strategies for drylands: available options and unanswered questions' in Mabbutt (ed.) 1979: 26–35.

Judson, S. (1968) 'Erosion of the land, or what's happening to our continents?', *Amer. Scientist*, **56**, 356–74.

Karmon, Y. (1971) *Israel: a regional geography*, Wiley, London.

Karmon, Y. (1976) 'Eilat, problems of a port on a desert coast', in Amiran and Ben-Arieh (eds) 1976: 106–37.

Kates, R. W. (1981) 'Drought impact in the Sahelian-Sudanic Zone of West Africa: a comparative analysis of 1910–15 and 1968–74', *Center for Technology, En-*

vironment, and Development, Background Paper No. 2, Clark University, Worcester, Mass.

Katz, R. W. and Glantz, M. H. (1979) 'Weather modification for food production: panacea or placebo? *J. Soil and Water Conservation*, **34**, 132–34.

Kellogg, W. W. (1978) 'Review of mankind's impact on global climate', in M. H. Glantz *et al.* (eds) *Multidisciplinary Research Related to the Atmospheric Sciences*, Nat. Centre Atmos. Res. Boulder, Col., 64–81.

Kelso, M. M. *et al.* (1973) *Water Supplies and Economic Growth in an Arid Environment*, University of Arizona, Tucson.

Kessler, E., Alexander, D. Y. and Rarick, J. F. (1978) 'Duststorms from the U.S. High Plains in late winter 1977 – search for cause and implications', *Proc. Okla. Acad. Sci.*, **58**, 116–28.

Kharin, N. G. and Petrov, M. P. (1975) *Glossary of Terms on Desert Environment and Land Reclamation*, Ylym Pub., Ashkhabad.

Khogali, M. M. (1979) 'Nomads and their sedentarization in the Sudan' in Mabbutt (ed.) 1979: 55–9.

King, R. (1972) 'The pilgrimage to Mecca: some geographical and historical aspects', *Erdkunde*, **26**, 61–73.

Kirkman, J. (1976) *City of Sanā*, British Museum, Ethnography Dept., London.

Kokot, D. F. (1955) 'Desert encroachment in South Africa', *Africa Soils*, **3**(3), 404–9.

Kolars, J. (1966) 'Locational aspects of cultural ecology: the case of the goat in non-western agriculture', *Geog. Rev.*, **56**, 577–84.

Kollmorgen, W. M. (1969) 'The Woodsman's assaults on the domain of the Cattlemen', *Annals Assoc. Amer. Geogs.*, **59**, 215–39.

Kollmorgen, W. M. and Jenks, G. F. (1958) 'Suitcase farming in Sully County, South Dakota', *Annals Assoc. Amer. Geogs.*, **48**, 27–40.

Köppen, W. (1931) *Die Klimate der Erde*, Berlin.

Kovda, V. A (1961) 'Land use development in the arid regions of the Russian plain, the Caucasus and central Asia' in Stamp (ed) 1961: 175–218.

Kraenzel, C. F. (1955) *The Great Plains in Transition*, University of Oklahoma, Norman.

Kramer, F. L. (1967) 'Eduard Hahn and the end of the "Three stages of man" ', *Geog. Rev.*, **57**, 73–89.

Lamb, H. (1963) *Genghis Khan: emperor of all men*, Bantam, New York (1st edn, 1927).

Lamb, H. H. (1977) *Climate: present, past and future*, Methuen, London, 2 vols.

Lapidus, I. M. (ed.) (1969) *Middle Eastern Cities*, University of California, Berkeley.

Lattimore, O. (1962) *Nomads and Commissars: Mongolia revisited*, Oxford University, New York.

Lawrence, R. (1969) *Aboriginal Habitat and Economy*, Aust. Nat. University, Dept. Geog. Occas. Paper 6, Canberra.

Lawrence, T. E. (1963) *Seven Pillars of Wisdom*, Dell, New York (1st edn, 1926).

Lee, D. H. K. (1964) 'Terrestrial animals in dry heat: man in the desert' in Field (ed.) 1964: 551–82.

Lee, D. H. K. (1969) 'Variability in human response to arid environments' in McGinnies and Goldman (eds.) 1969: 227–46.

Lee, D. M. (1979) 'Australian drought watch system' in M. T. Hinchley (ed.) *Proceedings of the Symposium on Drought in Botswana*, University Press of New

References

England, Hanover, USA, 173–208.
Lee, K. E. and Wood, T. G. (1971) *Termites and Soils*, Academic Press, London.
Lee, R. B. and De Vore, I. (eds) (1976) *Kalahari Hunter-Gatherers: studies of the Kung San and their neighbours*, Harvard University, Cambridge, Mass.
Leeds, A. and Vayda, A. P. (eds) (1965) *Man, Culture and Animals*, Amer. Assoc. Adv. Sci. Pub. No. 78, Washington.
Lhote, H. (1960) *The Search for the Tassili Frescoes* (trans. A. H. Brodrick), Readers Union, London.
Lloyd, C. (1974) *The Search for the Niger*, Readers Union, Newton Abbot.
Lobeck, A. K. (1939) *Geomorphology: an introduction to the study of landscapes*, McGraw-Hill, New York.
Logan, R. F. (1960) *The Central Namib Desert*, US Nat. Acad. Sci. Nat. Res. Council Pub. 758, Washington.
Logan, R. F. (1961a) 'Winter temperatures of a mid-latitude desert mountain range', *Geog. Rev.*, **51**, 236–52.
Logan, R. F. (1961b) 'Post-Columbian developments in the arid regions of the United States of America' in Stamp (ed.) 1961: 277–98.
Logan, R. F. (1969) *Bibliography of South West Africa: geography and related fields*, South West Africa Scientific Society, Windhoek.
Logan, R. F. (1973) 'The utilization of the Namib Desert, South West Africa' in Amiran and Wilson (eds) 1973: 177–86.
Lonsdale, R. E. and Holmes, J. H. (eds) (1981) *Settlement Systems in Sparsely Populated Regions, the United States and Australia*, Pergamon, New York.
Love, R. M. (1971) 'The rangelands of the western U.S' in Ehrlich (ed.) 1971: 229–35.
Lovett, J. V. (ed.) (1973) *The Environmental, Economic and Social Significance of Drought*, Angus and Robertson, Sydney.
Lowdermilk, W. C. (1953) 'Conquest of the land through 7,000 years', *U.S. Dept. Agric. Soil Cons. Service Agric. Info. Bull.* No. 99.
Lowe, C. H. (1968) 'Appraisal of research on fauna of desert environments' in McGinnies, Goldman and Paylore (eds), 1968: 569–648.
Lynn-Smith, T. (1969) 'Agricultural–pastoral conflict: a major obstacle in the process of rural development', *J. of Inter-American Studies*, **11**, 16–43.
Mabbutt, J. A. (1976) 'Physiographic setting as an indication of inherent resistance to desertification' in WGDAL 1976: 189–97.
Mabbutt, J. A. (1978) *Desertification in Australia*, Water Research Foundation of Australia Report No. 54, Kingsford.
Mabbutt, J. A. (ed.) (1979) *Proceedings of the Khartoum Workshop on Arid Lands Management*, United Nations University, Tokyo.
Macfarlane, W. V. (1968) 'Protein from the wasteland: water and the physiological ecology of ruminants', *Aust. J. Sci.*, **31**, 20–30.
Macfarlane, W. V., Morris, R. J. H. and Howard, B. (1963) 'Turnover and distribution of water in desert camels, sheep cattle and kangaroos', *Nature*, **197**, 270–1.
McGinnies, W. G. and Goldman, B. J. (eds) (1969) *Arid Lands in Perspective*, University of Arizona, Tucson.
McGinnies, W. G., Goldman, B. J. and Paylore, P. (eds) (1968) *Deserts of the World*, University of Arizona, Tucson.
McGinnies, W. G., Goldman, B. J. and Paylore, P. (eds). (1971) *Food, Fiber and*

the Arid Lands, University of Arizona, Tucson.

McKelvey, V. E. (1972) 'Mineral resource estimates and public policy', *Amer. Scientist*, **60**, 32–40.

McKnight, T. L. (1969) *The Camel in Australia*, Melbourne University Press, Carlton.

Manners, G. (1981) 'Our planet's resources', *Geog. J.*, **147**, 1–22.

Marbut, C. F. (1935) 'Soils of the United States', *Atlas of American Agriculture*, US Dept. Agric., Washington.

Marshall, A. (1963) 'The prediction of indoor heat discomfort', *Aust. Geog. Stud.*, **1**, 115–23.

Marshall, J. K. (1973) 'Drought, land use and soil erosion' in Lovett (ed.) 1973: 55–77.

Martin, P. S. (1967) 'Prehistoric overkill' in P. S. Martin and H. E. Wright (eds), *Pleistocene Extinctions*, Yale University, New Haven.

Martonne, E. de and Aufrère, L. (1927) 'Map of interior basin drainage', *Geog. Review*, **17** (3), 414.

Mather, E. C. (1972) 'The American Great Plains', *Annals Assoc. Amer. Geogs.*, **62**, 237–57.

Meigs, P. (1952) 'Arid and semiarid climatic types of the world', *Proc. Eighth General Assembly and Seventeenth International Congress, International Geographical Union*, Washington, 135–8.

Meigs, P. (1953) 'World distribution of arid and semi-arid homoclimates' in UNESCO *Arid Zone Res. Series No. 1, Arid Zone Hydrology*, 203–9.

Melamid, A. (1973) 'Satellization in Iranian crude-oil production', *Geog. Rev.*, **63**, 27–43.

Miller, W. H. (1968) 'Santa Ana winds and crime', *Prof. Geog.*, **20**, 23–7.

Mitchell, C. W. and Willimott, S. G. (1974) 'Dayas of the Moroccan Sahara and other arid regions', *Geog. J.*, **140**(3), 441–53.

Mollin, F. E. (1938) *If and When It Rains – the Stockman's View of the Range Question*, Amer. Nat. Livestock Assoc., Denver.

Monteil, V. (1959) 'The evolution and setting of the nomads of the Sahara', *Int. Soc. Sci. J.*, **11**, 572–85.

Morales, C. (1977) 'Rainfall variability – a natural phenomenon', *Ambio*, **6**, 30–3.

Morrell, W. P. (1940) *The Gold Rushes*, Black, London.

Morse, R. N. (1977) 'Solar energy in Australia', *Ambio*, **6**, 209–15.

Murphy, D. R. (1971) 'The role of changing external relations on the growth of Las Vegas, Nevada, USA', *Selected Papers 21st Int. Geog. Congress*, Calcutta, **3**, 193–8.

Murphy, R. E. (1968) 'Landform regions of the world', Map Supp. No. 9, *Annal Assoc. Amer. Geogs.*, **58**.

Nace, R. L. (1969) 'World water inventory and control', in Chorley *et al.* (eds) 1969: 31–42.

Nelson L. (1952) *The Mormon Village: a pattern and technique of land settlement*, University of Utah, Salt Lake City.

Newman, J. E. and Pickett, R. C. (1974) 'World climates and food supply variations', *Science*, **186**(4167), 877–81.

Nir, D. (1974) *The Semi-arid World: man on the fringe of the desert*, Longman, Harlow.

Oberholzer, S. F. (1957) 'An economic investigation of farming in two extensive cattle-farming regions of the Union of South Africa', *South Africa Dept. Agric. Bull.*, **350**, Cape Town.

Odum, E. P. (1963) *Ecology*, Holt, Rinehart, and Winston, New York.

Overton, J. D. (1981) 'A theory of exploration', *J. Hist. Geog.*, **7**, 53–70.

Ovington, J. D. *et al.* (1973) *A Study of the Impact of Tourism at Ayers Rock-Mt. Olga National Park*, Australian Govt. Pub. Ser., Canberra.

Owen, D. L. (1981) *Providence their Guide: the Long Range Desert Group, 1940–5*, Harrap, London.

Parliament of the Commonwealth of Australia (1977) *Solar Energy: Report of the Senate Standing Committee on National Resources*, Australian Govt. Pub. Ser., Canberra.

Parry, J. H. (1963) *The Age of Reconnaissance*, Weidenfeld and Nicolson, London.

Paulik, G. J. (1971) 'Anchovies, birds and fishermen' in W. W. Murdoch (ed.) 1971 *Environment: resources, pollution and society*, Sinauer, Stamford, Conn., 156–85.

Paylore, P. (1966) *Seventy-five Years of Arid Lands Research at the University of Arizona*, University of Arizona, Tucson.

Paylore, P. (ed.) (1976) *Desertification: a world bibliography*, University of Arizona, Tucson.

Paylore, P. and Greenwell, J. R. (1979) 'Fools rush in: pinpointing the Arid Zones', *Arid Lands Newsletter* (Arizona), **10**, 17–18.

Paylore, P. and Greenwell, J. R. (1980) 'Fools rush in. Part 2: selected arid lands population data', *Arid Lands Newsletter*, **12**, 14–18.

Paylore, P. and McGinnies, W. G. (1969) *Desert Research: selected references 1965–68*, US Army, Natick, Mass.

Peet, J. R. (1969) 'The spatial expansion of commercial agriculture in the nineteenth century: a Von Thünen interpretation', *Econ. Geog.*, **45**, 283–301.

Penck, A. (1894) *Morphologie der Erdoberfläche*, Stuttgart.

Peppelenbosch, P. G. N. (1968) 'Nomadism on the Arabian Peninsula: a general appraisal', *Tijd. Econ. Soc. Geog.*, **59**, 335–46.

Perren, R. (1971) 'The North American beef and cattle trade with Great Britain, 1870–1914', *Econ. Hist. Rev.*, 2nd Series, **24**, 430–44.

Peterson, D. F. and Crawford, A. B. (eds) (1978) *Values and Choices in the Development of the Colorado River Basin*, University of Arizona, Tucson.

Petrov, M. (1976) *Deserts of the World* (trans. R. Lavott), Wiley, New York.

Pierce, R. A. (1960) *Russian Central Asia 1867–1917: a study in colonial rule*, University of California, Berkeley.

Pollard, W. G. (1976) 'The long range prospects for solar energy', *Amer. Scientist*, **64**, 424–9.

Powell, J. M. and Williams, M. (eds) (1975) *Australian Space: Australian Time: geographical perspectives*, Oxford University, Melbourne.

Rabbitt, M. C. (1979) *Minerals, Lands, and Geology for the Common Defence and General Welfare*, Vol. 1. *Before 1879*, US Govt. Print. Office, Washington.

Rainey, R. C. (1977) 'Rainfall: scarce resource in "opportunity country"', *Phil. Trans. R. Soc. London*, B **278**, 439–55.

Ratcliffe, F. (1963) *Flying Fox and Drifting Sand*, Angus and Robertson, Sydney (1st edn, 1938).

Rees, R. (1969) 'The small towns of Saskatchewan', *Landscape*, **18**(3), 29–33.

Regan, D. L. (1980) 'Marine biotechnology and the use of arid zones', *Search*, **11** (11), 377–81.

Reifenberg, A. (1955) *The Struggle Between the Desert and the Sown*, Jewish Agency, Jerusalem.

Revelle, R. (1971) 'Water' in Ehrlich (ed.) 1971: 56–67.

Roberts, G. E. and Sheridan, N. R. (1969) *Air Conditioned Housing for Northern Australia*, University of Queensland Solar Res. Notes No. 3, Brisbane.

Rogers, J. A. (1981) 'Fools rush in, Part 3: selected dryland areas of the world', *Arid Lands Newsletter* (Arizona), **14**, 24–5.

Rowley, C. D. (1972) *The Destruction of Aboriginal Society*, Penguin, Harmondsworth (1st edn. 1970).

Rutherford, J. (1964) 'Interplay of American and Australian ideas for development of water projects in northern Victoria', *Annals Assoc. Amer. Geogs.*, **54**: 88–106.

Ruttan, V. W. (1965) *The Economic Demand for Irrigated Acreage: new methodology and some preliminary projections 1954–1980*, Johns Hopkins, Baltimore.

Saini, B. S. (1962) 'Housing in the hot arid tropics', *Architectural Science Review*, **5**, 3–12.

Saini, B. S. (1970) *Architecture in Tropical Australia*, Melbourne University, Carlton.

Sandars, N. K. (1960) *The Epic of Gilgamesh*, Penguin, Harmondsworth.

Sauer, C. O. (1952) *Agricultural Origins and Dispersals*, Amer. Geog. Soc., New York.

Schmidt-Nielsen, K. (1956) 'Animals and arid conditions: physiological aspects of productivity and management' in G. F. White (ed.) *The Future of Arid Lands*, Amer. Assoc. Adv. Sci., Pub. No. 43, Washington.

Schmidt-Nielsen, K. (1965) *Desert Animals: physiological problems of heat and water*, Oxford.

Schneider, W. (1963) *Babylon is Everywhere*, Hodder and Stoughton, London.

Schulz, J. E. (1967) 'Water in the soil and how to keep it there', *South Australia J. of Agric.*, **71**, 76–83.

Sears, P. B. (1949) *Deserts on the March*, Routledge and Kegan Paul, London.

Shantz, H. L. (1956) 'History and problems of arid lands development' in G. F. White (ed.) *The Future of Arid Lands*. Amer. Assoc. Adv. Sci. Pub. No. 43, Washington.

Sharon, D. (1979) 'Studies on the spatial structure of desert rainfall', unpub.

Shibl, Y. (1971) *The Aswan High Dam*, Arab Inst. Res. Pubn., Beirut.

Shreiber, J. F. and Matlock, W. G. (1978) *The Phosphate Rock Industry in North and West Africa*, University of Arizona, Tucson.

Silberman, L. (1959) 'The Somali Nomads', *Int. Soc. Sci. J.* **11**, 559–71.

Simoons, F. (1967) *Eat Not This Flesh: food avoidances in the Old World*, University of Wisconsin, Milwaukee.

Singer, C. S. *et al.* (eds) (1956) *A History of Technology*, 2 vols, Oxford University, London.

Skinner, B. J. (1976) 'A second iron age ahead?', *Amer. Scientist*, **64**, 258–69.

Slatyer, R. O. and Perry R. A. (eds) (1969) *Arid Lands of Australia*, Australian National University, Canberra.

Solbrig, O. T. and Orians, G. H. (1977) 'The adaptive characteristics of desert plants', *Amer. Scientist*, **65**, 412–21.

Spate, O. H. K. (1953) *The Compass of Geography*, Australian National University, Canberra.

References

Spence, C. C. (1980) *The Rainmakers: American 'pluviculture' to World War II*, University of Nebraska, Lincoln.

Stamp, L. D. (ed.) (1961) *A History of Land Use in Arid Regions*, UNESCO Arid Zone Res. No. 17, Paris.

Strickon, A. (1965) 'The Euro-American ranching complex' in Leeds and Vayda (eds.) 1965: 229–58.

Sweet, L. E. (1965) 'Camel pastoralism in north Arabia and the minimal camping unit' in Leeds and Vayda (eds) 1965: 129–52.

Talbot, W. J. (1961) 'Land utilization in the arid regions of southern Africa. Part I. South Africa' in Stamp (ed.) 1961: 299–31.

Templer, O. W. (1978) *The Geography of Arid Lands: a basic bibliography*, Int. Center for Arid and Semi-arid Land Studies, Texas Tech. University, Lubbock.

Thesiger, W. (1960) *Arabian Sands*, Readers Union, London.

Thimm, H-U. (1979) 'Socio-economic assessment of agricultural development projects in the Sudan' in Mabbutt (ed.) 1979: 71–5.

Thomas, B. (1932) *Arabia Felix: across the 'Empty Quarter' of Arabia*, Scribner's, New York.

Thompson, J. W. (1942) *A History of Livestock Raising in the United States, 1607–1860*, US Bur. Agric. Econ., Agric. Hist. Series No. 5, Washington.

Thorne, W. (ed.) (1963) *Land and Water Use*, Amer. Assoc. Adv. Sci. Pub. No. 73, Washington.

Thornthwaite, C. W. (1948) 'An approach towards a rational classification of climate', *Geog. Rev.*, **38**, 55–94.

Tinus R. W. (ed.) (1976) *Shelterbelts on the Plains*, Gt. Plains Agric. Council Pub. No. 78, Denver.

Tobias, P. V. (1964) 'Bushmen hunter–gatherers: a study in human ecology' in Davis (ed.) 1964: 67–86.

Toennies, F. (1957) *Community and Society (Gemeinschaft und Gesellschaft)*, (trans. by C. P. Loomis), Harper, New York (1st edn, 1887).

Twidale, C. R. (1981) 'Granitic inselbergs: domed, block-strewn and castellated', *Geog. J.*, **147**, 54–71.

Ucko, P. J. and Dimbleby, G. W. (eds) (1971) *The Domestication and Exploitation of Plants and Animals*, Duckworth, London.

UN (1977a) *World Map of Desertification at a scale of 1 : 25 000 000*, United Nations Conference on Desertification (+ Commentary).

UN (1977b) *United Nations Conference on Desertification: round-up, plan of action and resolutions*, United Nations, New York.

UN (1977c) *World Distribution of Arid Regions*, scale 1:25 000 000, CERCG, Paris.

UN Secretariat (1977) *Desertification: its causes and consequences*, Pergamon Press, Oxford.

University of Adelaide (1973) *The Change and Challenge of our Arid Lands*, Water Research Foundation Rep. No. 40, Adelaide.

Updike, J. (1979) *The Coup*, Penguin, Harmondsworth.

USA (1936) *The Western Range*, US 74th Congress, 2nd Session, Doc. No. 199, Washington.

USA (1970) *One Third of the Nation's Land: a report to the President and Congress by the Public Land Law Review Commission*, Govt. Printing Office, Washington.

Vallentine, H. R. (1967) *Water in the Service of Man*, Penguin, Harmondsworth.

Vita-Finzi, C. (1971) 'Geological opportunism' in Ucko and Dimbleby (eds) 1971: 31–4.

Wagner, P. L. and Mikesell, M. W. (1962) *Readings in Cultural Geography*, University of Chicago, Chicago.

Walls, J. (1980) *Land, Man and Sand; desertification and its solution*, Macmillan, New York.

Ward, D. (1964) 'A comparative historical geography of streetcar suburbs in Boston, Massachusetts and Leeds, England: 1850–1920', *Annals Assoc. Amer. Geogrs.*, **54**, 477–89.

Warrick, R. A. (1975) *Drought Hazard in the United States: a research assessment*, University of Colorado, Boulder.

Washburn, S. L. (1976) 'Foreword' in Lee and De Vore (eds) 1976.

Watson, J. A. L. and Gray, F. J. (1970) 'The role of grass-eating termites in the degradation of a Mulga ecosystem', *Search*, **1**, 43.

Watt, K. E. F. (1973) *Principles of Environmental Science*, McGraw-Hill, New York.

Weihe, D. L. *et al.* (1979) 'Guayule as a commercial source of natural rubber' in Goodin and Northington (eds) 1979: 230–43.

Weissleder, W. (ed.) (1978) *The Nomadic Alternative*, Mouton, Hague.

Wellard, J. (1973) *By the Waters of Babylon*, Readers Union, Newton Abbot.

Went, F. W. (1955) 'The ecology of desert plants', *Sci. American*, **192**(4), 68–75.

WGDAL (1976) *Problems in the Development and Conservation of Desert and Semi-desert Lands*, Symposium K26, Ashkhabad USSR, Working Group on Desertification in and around Arid Lands, University of New South Wales, Sydney.

Wheatley, P. (1971) *The Pivot of the Four Quarters: a preliminary enquiry into the origins and character of the ancient Chinese city*, Edinburgh University, Edinburgh.

White, G. F. (ed.) (1956) *The Future of Arid Lands*, Amer. Assoc. Adv. Sci. Pub. No. 43, Washington.

White, G. F. (1960) *Science and the Future of Arid Lands*, UNESCO, Paris.

White, G. F. (ed.) (1974) *Natural Hazards: local, national, global*, Oxford University, New York.

White, G. F. (ed.) (1978) *Environmental Effects of Arid Land Irrigation in Developing Countries*, MAB Tech. Note No. 8, UNESCO, Paris.

Whorf, B. L. (1956) *Language, Thought and Reality*, Harvard University, Cambridge, Mass.

Whyte, R. O (1961) 'Evolution of land use in south-western Asia' in Stamp (ed.) 1961: 57–118.

Wiener, A. (1977) 'Coping with water deficiency in arid and semi-arid countries through high-efficiency water management', *Ambio*, **6**, 77–82.

Wiens, H. J. (1969) 'Change in the ethnography and land use of the Ili Valley and Region, Chinese Turkestan', *Annals Assoc. Amer. Geogs.*, **59**, 753–775.

Williams, M. (1974) *The Making of the South Australian Landscape*, Academic Press, London.

Williams, M. (1975) 'More and smaller is better: Australian rural settlement 1788–1914' in Powell and Williams (eds) 1975: 61–103.

Wilson, A. W. (1973) 'The larger urban centers of the coastal deserts' in Amiran and Wilson (eds) 1973: 33–6.

Wilson, A. W. (1980) 'Tucson – the city and the desert' in J. A. Mabbutt and A. W. Wilson (eds) (1980) *Social and Environmental Aspects of Desertification*, United Nations University, Tokyo, 3–4.

315

References

Winters, C. (1977) 'Traditional urbanism in the north central Sudan', *Annals Assocn. Amer. Geogs.*, **67**, 500–20.

Wittfogel, K. A. (1970) *Oriental Despotism: a comparative study of total power*, Yale University, New Haven (1st edn 1957).

Wollman, N. (ed.) (1962) *The Value of Water in Alternative Uses*, University of New Mexico, Albuquerque.

Woods, A. (1977) 'Wind and water power', *Current Affairs Bulletin* (Sydney), **54**, 22–7.

Young, R. S. (1976) 'Viking on Mars: a preliminary survey', *Amer. Scientist*, **64**, 620–7.

Zahran, M. A. *et al.* (1979) 'Economic potentialities of *Juncus* plants' in Goodin and Northington 1979: 244–60.

Zeuner, F. E. (1963) *A History of Domesticated Animals*, Hutchinson, London.

Index

Index

Index